U.S.NRC

United States Nuclear Regulatory Commission

Protecting People and the Environment

NUREG-1964

I0482705

Access Control Systems

Technical Information for NRC Licensees

Office of Nuclear Security and Incident Response

AVAILABILITY OF REFERENCE MATERIALS
IN NRC PUBLICATIONS

United States Nuclear Regulatory Commission

Protecting People and the Environment

NUREG-1964

Access Control Systems

Technical Information for NRC Licensees

Manuscript Completed: March 2011
Date Published: April 2011

Office of Nuclear Security and Incident Response

ABSTRACT

This report provides technical details applicable to access control methods and technologies commonly used to protect facilities licensed by the U.S. Nuclear Regulatory Commission. It contains information on the application, use, function, installation, maintenance, and testing parameters for access control and search equipment and the implementation of protective measures that support access control. This information is intended to assist licensees in designing, installing, employing, and maintaining access control systems at their facilities.

CONTENTS

LIST OF FIGURES

LIST OF TABLES

1. INTRODUCTION

1.1 Overview

This report provides information relative to designing, installing, testing, maintaining, and monitoring access control systems used for the protection of facilities licensed by the U.S. Nuclear Regulatory Commission (NRC). The regulations in Title 10 of the *Code of Federal Regulations* (10 CFR) Part 73, "Physical Protection of Plants and Materials" address the NRC's requirements for the physical protection of nuclear power reactor facilities, independent spent fuel storage facilities, fuel cycle facilities, strategic special nuclear materials, and special nuclear material (SNM) of moderate and low strategic significance. The type of NRC licensee and the SNM authorized to be possessed by that licensee determine the specific requirements for physical protection programs at NRC-licensed facilities.

Access control is a significant element of any physical protection program. The overall objective of access control is to ensure that only authorized and properly searched personnel, vehicles, and materials are granted access to, and exit from, areas that require protection. To ensure effectiveness, the physical protection program should integrate access control measures that complement the implementation of other components, such as physical barriers, intrusion detection and assessment systems, and search processes. Site-specific conditions should also be analyzed and considered when establishing and implementing measures and technologies to control access.

This report provides technical details pertaining to access control measures and technologies commonly used in physical protection programs requiring moderate- to high-level security applications. The following information is provided for consideration when designing an access control system and provides a broad overview of available access control equipment and specialized detection devices.

1.2 Design Considerations

To effectively control access, the areas to be protected must first be identified. Once these areas are identified, the use of physical barriers at the perimeter of the identified areas should be considered to prevent or delay unauthorized access and to provide a means to detect and observe attempted or actual unauthorized access by personnel, vehicles, and materials beyond these physical barriers. To support facility operations and satisfy access control requirements, physical barrier systems should be designed to channel the movement of personnel, vehicles, and materials to designated locations at which such personnel, vehicles, and materials will be processed before granting access. This report refers generically to these locations as "access control portals."

Access control portals can provide a broad variety of functions or may be limited in scope, depending on the needs of the facility and the nature of the processing to be accomplished. The specific process to be used and criteria to be met at these locations will be based on applicable requirements, the needs of the facility, and site-specific conditions. Typically, an access control portal will have the four following characteristics:

(1) a location and process where identity and authorization for access can be verified and identification media can be issued

(2) search equipment and search personnel, to verify that unauthorized items are not present prior to entry

(3) alarmed entry control devices (e.g., doors, gates, turnstiles, card readers, biometrics) that prevent or delay, detect and observe actual or attempted unauthorized entry before completion of the required processing

(4) an individual who controls the last access control function and possesses the capability to deny access and call for assistance if needed

The design of access control portals should consider the layout and arrangement of search equipment, search personnel, and access control devices. Alternate access portals to accommodate the ingress of larger items may be used, provided appropriate controls for access are followed. Section 4, "Schematic Drawings of Access Control Portals," provides information on layout and arrangement of personnel access control portals.

An effective access control process should do the following:

- Verify the identity of personnel, vehicles, and materials.

- Verify the authorization for access of personnel, vehicles, and materials requesting access.

- Ensure that personnel, vehicles, and materials being considered for access do not possess unauthorized items.

- Be designed and configured to prevent or delay and detect and observe attempted or actual bypass.

When designing an access control system, considerations should include the following:

- The type of area (security level of facility) into which access is being controlled

- Facility layout, with specific consideration given to the configuration of the access control facility, search equipment and trained and qualified search personnel (search train), and the physical barriers that support access control

- The desired rate at which personnel, vehicles, and materials are expected to move through the access control processes (throughput rate)

- Search equipment and trained and qualified search personnel, to ensure that unauthorized items are not introduced into the facility (selection of equipment and the training and qualifications of personnel should be consistent with the physical protection program design goals of the facility)

- Countermeasures for the capability and methodologies to defeat or bypass the search train, physical barriers, and access control measures

- The amount of attendant personnel required at access control portals to ensure effective access controls

- Secondary power for security equipment and systems to ensure that access control systems remain operable and available for use

Section 2, "Discussion of Technologies," describes a variety of technologies used at access control portals to control access, and Section 4, "Schematic Drawings of Access Control ," provides technical details relative to the design of access control portals to maximize the effectiveness of search trains.

Access control portals should be designed to provide defense in depth through the integration of various systems, technologies, equipment, and supporting processes, as needed, to effectively control access consistent with the physical protection program design goals established for the facility being protected. Access control measures that incorporate the integration of diverse and redundant equipment and systems, also provide flexibility when equipment fails. In the majority of cases, the implementation of diverse technologies provides a more comprehensive access control and unauthorized item detection capability than most single technologies which have some characteristic weak points.

One of the first considerations for the implementation of access controls is verifying the identity of personnel attempting access and confirming that they are authorized to enter the facility. The identity of personnel and their authorization for access can be verified in many ways. The identity of personnel is most commonly verified by attendant personnel or equipment through something possessed by an individual (e.g., driver's license, photo badge, keycard), something about the individual (e.g., facial features, fingerprint or palm print, or other biometrics), or something known to an individual (personal identification number (PIN) or passcode). By using a combination of these three criteria, positive identification of an individual may be obtained.

One common method for identifying personnel who are authorized access to a facility's protected areas is the use of photo identification badges. Most badging systems allow the assignment of access privileges to the badge holder through electronic codes contained within each badge. Personnel requesting entry to the protected area of a facility are usually required to present their badge to a member of the security organization located at the access control portal. To confirm the individual's identity, the member of the security organization should compare the photo on the badge to the individual's facial characteristics. Methodologies to verify badge authenticity, such as visual and physical inspection, should also be implemented. Verification of the badge as a certified credential is essential.

Aside from touching the badge, the security officer can recognize other visual and physical characteristics of the badge to assist in verifying the credential as authentic. Additionally, the authenticity of the badge can be verified through the use of an electronic database that associates the badge with a biometric feature of the individual such that both are needed before entry. Automated access control devices may be used to verify a pre-enrolled biometric characteristic of the individual in conjunction with the electronic coding assigned to each badge. Alternatively, a PIN could be used in conjunction with, or in lieu of, a biometric feature in the automated access control system scheme. In this methodology, the person requesting access would type the PIN into a key pad at the badge access control station in conjunction with a badge swipe or biometric identification. Use of all three methodologies provides defense in depth by combining who a person is (i.e., biometric characteristics) with what the person has (i.e., electronic badge) and what the person knows (i.e., PIN).

Section 2.1.3, "Biometric Devices," provides technical guidance for biometric devices. These devices should be maintained to have the correct false accept/false reject rate. See Section 2.1 "Identity Verification," for more detailed discussion regarding the picture badge/keycard and biometric systems. Section 3, "Testing of Search Equipment and Methodologies," provides guidance for establishing the correct false accept/false reject rate. Automated access control systems should include antipassback (prevention of an individual's re-entry without logging out of the access control system) and antipiggyback (prevention of multiple individuals entering using a single authorized credential) features.

Site procedures should address actions to be taken upon discovery of nonfunctioning access control systems. Compensatory measures or backup equipment should be utilized in the case of nonfunctioning systems. Training of security personnel may assist in the detection of equipment performance degradation and the potential for bypass.

In addition to ensuring that personnel, vehicles, and materials are identified and authorized for access, a process for searching personnel, vehicles, and materials to confirm that they do not contain or possess unauthorized items should be implemented. A typical search process can be divided into four major categories: (1) search area, (2) search equipment, (3) personnel conducting the search and search process oversight, and (4) search procedures. The search process is normally accomplished in predetermined locations using visual observation by attendant security personnel and search equipment in accordance with site procedures. Search equipment normally includes metal detectors, explosive detectors, and X-Ray devices, and may include radiation detectors. When designing an access control system, the selection of search equipment should consider the site-specific configuration and items that must be excluded from entry or exit. Section 2 of this report provides additional information relative to these technologies. Search equipment alarms should result in an immediate response by attendant security personnel, which should include a temporary denial of access pending the completion of alarm resolution, which usually includes a physical pat-down search to confirm or exclude the presence of unauthorized items. For search processes to be effective, the individuals performing search and oversight activities must be trained and qualified in search processes and remain alert, attentive, and vigilant at all times.

When designing access control portals, the layout and arrangement of the portal area and the search train should provide enough capacity to satisfy required access control processes and enable the attendant security personnel to maintain observation and control of personnel, materials, and vehicles entering the facility's protected area. The layout and arrangement of access control portals and search train equipment should also consider the desired throughput rate. Factors that should be considered when designing an access control portal to accommodate desired throughput rates should include the number of search trains and the number of attendant security personnel needed to perform required access control processes during periods of increased or peak personnel, material, and vehicle traffic. Optimally, the throughput rate should support the ingress of personnel, materials, and vehicles consistent with operational needs without degrading the level of security required at the facility.

Each access control portal should include a method to control the area for processing personnel and their possessions. This can be accomplished through the use of physical barriers configured to maintain organization and flow and to ensure the separation of personnel entering the facility's protected area from other personnel exiting the area. Properly designed access control portals do not permit exiting personnel to reenter the facility's protected area without satisfying the access control and search processes before each reentry. Other measures, such as the incorporation of intrusion detection and assessment equipment, should be considered to support access control and the capability to detect, assess, and initiate response to attempted unauthorized bypass. Intrusion detection devices should be affixed to all access control portal barriers, and video assessment equipment should be located in the vicinity of the portal barrier and should be oriented to view the barriers and the area leading to the barriers. Intrusion detection and video assessment equipment at access control portals should annunciate and display in the facility's alarm stations and in the location at which final access to the protected area is controlled.

The effects of environmental conditions on search equipment technology should be considered, especially if screening is performed outside. Some trace chemical sensors (e.g., ion mobility spectrometers (IMS)) are sensitive to changes in atmospheric pressure, temperature, and humidity, and may not function properly or may require recalibration if these conditions vary outside of defined ranges. Explosive-sniffing canines are also likely to perform better under some environmental conditions and not as well under others, such as extreme heat. When obtaining a piece of security equipment, the vendor should be consulted to determine the conditions under which the system can operate and to understand when the system may need to be recalibrated or simply replaced with an alternative means of detection, such as physical search. Appropriate calibration tests should be defined and implemented to determine that equipment is operating properly before it is used. Section 3, "Testing of Search Equipment and Methodologies," offers guidance on maintenance and testing. Search equipment should be tested and maintained on a regular basis to ensure that the equipment is functioning and accurate within required parameters.

1.3 Changes in Technology and Methodologies

As physical protection technologies and methodologies continue to evolve, the adversary's knowledge and capabilities to defeat these technologies and methodologies also evolves. As with any element of a physical protection program, an awareness of access control and search equipment capabilities and potential vulnerabilities should be maintained to ensure that measures to account for performance capability limitations and potential vulnerabilities can be implemented. Maintaining awareness in this area is especially critical for search equipment as evidenced by the increased amount of composite materials used in the manufacturing of commercially available firearms resulting in a decreased metal content and the combination and use of various commercially available compounds and chemicals to create undetectable improvised explosive devices. When search equipment capabilities are surpassed by technological advancements in the manufacturing and use of firearms and improvised explosive devices, physical and visual searches should be implemented to account for the specific search equipment limitations.

2. DISCUSSION OF TECHNOLOGIES

This section provides background information on the various technologies discussed in this report and introduces the key features of techniques for access control for personnel, package, and vehicle screening applications.

For additional information, see the references listed at the end of this report.

2.1 Identity Verification

Verification of identity is performed by some combination of the following three factors:

- (1)   Something the person requesting access has in their possession

- (2)   Some characteristic of the person that can be measured

- (3)   Something the person knows

In most access control systems, a coded credential serves as something in the user's possession. Verifying that the correct person is carrying that credential can be accomplished using one or both of the other two factors. The measurements of user characteristics are performed using biometric devices. The third factor, something the person knows, usually takes the form of a PIN.

2.1.1 Card Reader Systems

Card reader systems are suitable for both interior and exterior physical access control applications to gates, doors, and other points of entry. Card reader systems provide an excellent platform to fulfill physical access control requirements and serve as a flexible foundation for customized security applications. Card readers may be used in combination with numeric keypads for PIN entry and provisions for connecting biometric devices to the system. In a card reader system that uses two-factor authentication, a biometric template or record of a PIN must be maintained at the access control panel or be provided to the panel when required through a computer network. Access control panels have a range of input and output capabilities. The panel also tracks the door status, open or closed, and whether the door is being forced or attacked. In some systems, the panel can generate an audible alarm for trouble conditions. The requirements for a point of entry either to "fail safe" or "fail secure," locked or unlocked respectively, should be carefully considered and discussed with the vendor before installing any new system.

Magnetic stripe cards can be used for authentication and access control in many security areas. Magnetic stripe card readers can be interfaced with a wide variety of access control equipment, including locally controlled electrical locks, access control panels, or centrally controlled security databases. Cards are encoded when they are issued. Magnetic stripe security cards can double as photo identification badges or keycards when they are printed or embossed with the user's name, identification number, imprints of corporate or organizational logos, and a photograph. In magnetic stripe systems, the card must be either "swiped" through a reader track or fully inserted into, and then removed from, a reader in order to cycle a lock. Card readers can be mounted to adjacent walls, doorframes, or any convenient horizontal surface. Depending on the system, the control of the lock can be located at the door, at an intermediate control panel, or at an alarm station to manage authentication processes for a large facility with many access points and authorized users. Systems can be programmed so that an alarm is generated after a preset number of failed access attempts. Magnetic stripe systems offer both fine-grained control of access points and detailed, real-time recordkeeping capability. Fine-grained access control is the ability to permit or deny access through a single door or set of doors. Magnetic stripe systems achieve this using centralized system control. In centralized systems, a computerized database is maintained which records authorized cardholders and their access permissions. When a cardholder operates a card reader, the reader sends a signal back to the controller, which identifies the cardbearer and access point. If a cardbearer has permission for this door, then the access point opens. Another advantage of centralized control is that the list of permissions for each user can change according to the situation.

Wiegand type access control systems consist of a card, a card reader, and an access control panel. Other functional elements may be present in some of these systems. Wiegand type access control systems were developed in the early 1970s and are used throughout the public and private sectors. The Wiegand type card is difficult to forge or counterfeit and is a crucial component of the system's dependability. A Wiegand type card is an example of a plastic or vinyl credit-card-sized device that carries two rows of small parallel wires of proprietary composition that generate a binary number when the card is passed through a reader. The wires are embedded within the card during the manufacturing process and provide the protection against forgery. Outside the reader's magnetic field, the embedded wires are essentially inert. Cards can be personalized with a photograph and identification information using a card printer.

Wiegand type cards are operated by swiping the card through the slot on a card reader. In the Wiegand type access control system, the card reader generates a magnetic field, which interacts with the wire segments that are embedded in the card. As the wires pass through the magnetic field, each wire produces a momentary pulse that is detected by the reader. The series of pulses forms a binary number that is transmitted from the card reader to the access control panel. If the binary number is a valid access control code, then the access control panel unlocks the point of entry. When new equipment is integrated into an existing Wiegand type access control system, the output of the reader device must be compatible with the other elements of the system, particularly the access control panel.

2.1.1.1 Contact Smart Cards

Contact smart card access control systems can be used in high-security applications involving large numbers of people that require significant flexibility and layering, but need only moderate throughput rates. Because contact smart cards must be inserted into and withdrawn from a reader, the system throughput tends to be lower than for contactless card systems. Contact smart cards are much more capable than keycard systems and magnetic stripe systems. Smart cards usually look like a common credit card or bank card. The type of embedded integrated circuit determines the capabilities of a card.

To operate a contact smart card system, a user must insert a card into a reader. The card reader provides power to the card, which causes the card to initialize. Depending on the type of card system, after initialization, the card can provide stored digital files, such as biometric templates or encryption keys, or perform algorithmic functions, which the security application programmers have defined. Regardless of the system architecture, if the card is authenticated, then the card reader sends a message to the access control panel to open the entry point. A multitude of card readers are commercially available. Contact smart card readers often have built in numeric keypads for entering PINs and provisions for connecting biometric devices for two-factor authentication.

2.1.1.2 Contactless Smart Cards

Some access control systems use smart cards that, in addition to having processing or memory capabilities, have a radiofrequency (RF) communications capability which allows a card reader to interact with the smart card at a distance. These cards are referred to as "contactless" smart cards. Contactless smart card systems can be used in situations that may require greater throughput rates than contact smart card systems. Some smart cards can be equipped with magnetic stripes, bar codes, and other systems to facilitate control and interact with other types of access systems with only the one device. The memory capability of a contactless smart card is generally no more than 9 kilobytes of data.

Smart cards resemble a common bank or credit card with an embedded microchip, but other forms are available. Contactless smart cards use RF transmitters, and an antenna for the contactless smart card is embedded alongside the microchip. As an individual approaches the entrance to a controlled area, the contactless smart card enters the detection field of the card reader. Usually, the card bearer passes the card in front of the card reader at a distance of no more than 6 inches, but readers with ranges of up to 6 feet are available.

Punching holes in the card to attach a lanyard or a key chain can sever portions of the card's electronics and ruin it. Strong magnetic fields can scramble or erase the memory on some cards. Cardholders should be instructed on the proper care of the card to reduce the incidence of these problems.

2.1.2 Key Fobs

Key fobs are used as interface devices to many access control and security systems. They can be made of plastic, epoxy resin, or metal, and are normally smaller than a credit card or business card.

Depending on the system that they control, key fobs can operate by a number of different methods. Garage door openers, facility alarm controllers, and fobs used in the automotive and transportation industries transmit a coded RF signal to a receiver inside the vehicle or facility when the operator presses an appropriate key or button. Key fobs using RF transmissions are usually made of plastic or epoxy resin to avoid interference with the transmitted signal. The radio signal contains a binary number that is a control code. Encrypting the transmitted signal can enhance the security of the system. Key fobs that do not use RF signals but use smart card, Wiegand type cards, or another technology require a compatible reader.

One advantage of the key fob is that it can be attached to a key ring, something that people take care not to lose. The key fob interface is suitable for controlling both interior and exterior applications.

2.1.3 Biometric Devices

The term "biometrics" refers to systems that can measure specific physiological or behavioral characteristics of a person to aid in the process of identity verification. Biometric access control refers to the use of human biological attributes for verification or identification in physical or logical access control systems. Biometric systems use physiological or behavioral data measurements to compare with previously enrolled information to determine such system responses as granting access or establishing identity. Biometric devices can be based upon any measurement of a human being. For the best performance, a biometric device should be based on a characteristic that (1) is distinct, (2) does not vary over time, and (3) is easy to collect.

Most access control systems feature physiological data (i.e., data that directly relate to the measurement of some aspect of the body's dimensions) or behavioral data (i.e., data that relate to the measurement of a body's reaction or performance) or both.

Biometric access control systems are automated so that they work without direct human intervention. They generally produce an access control decision in a few seconds or less. Any unique human physiological or behavioral characteristic can be used as a biometric identifier as long as it meets certain tests, as described below:

- Universality: Every person should possess it (excluding loss of limbs or birth defects in a particular individual).

- Distinctiveness: Any two persons should always exhibit different versions of the same trait.

- Permanence: The biometric should be confirmable over a long period of time.

- Collectability: It should be easily and quickly measurable.

A variety of factors need to be considered when choosing a biometric system, including the following:

- User acceptability

- Length of operating time

- Length of enrollment time

- Operating environment

- Database size

- Accuracy

Commonly available biometric systems are based on hand geometry, fingerprints, facial image, voice patterns, and iris patterns.

Before using a biometric device to verify the identity of an individual, the individual must be enrolled in the system. During this step, the system measures the biometric feature of the individual and a template is constructed for future comparison. This step requires a few seconds to several minutes depending on the technology used. Biometric access control devices extract measurements of the characteristic of interest, construct a mathematical template from those measurements, and compare this template to templates in an enrollment database to control access. In most cases, the entire picture or fingerprint is not used for comparison, just the template data. Privacy concerns are minimized, because the biometric parameter that was measured cannot be recreated from the template data. When implementing a biometric system, the following considerations apply:

- An enrollment station requires adequate space and resources.

- Templates created during enrollment should be protected and communicated to the access control system in a secure manner.

To gain access to a facility, the individual presents the physical feature (e.g., hand or face) to be measured. The system compares this measurement against the template created during enrollment. Typically, a score relating to the quality of the match is determined and compared against a threshold. If the score is above the threshold, it is judged a match and the user is granted access. If the score is below the threshold, the user is denied access. The accuracy of a biometric device is usually described by its false accept and false reject rates as defined below:

- The false accept rate is the rate at which an imposter is improperly authenticated as matching a template that is based on another person's biometric feature.

- The false reject rate is the rate at which authentic users will fail to match their own templates.

Both the false accept and false reject rates should be tested when first implementing a biometric system to ensure that the system meets security performance requirements. A program should also be in place to ensure that the system continues to function as intended and that regular maintenance and calibration are performed as required. Commercially available biometric systems normally operate with false accept and false reject rates of less than 1 percent.

In access control security applications, biometrics are used for either verification or identification. One of two questions must be answered: "Am I who I say I am?" (verification) or "Who am I?" (identification). Unlike passwords or PINs, the biometric enrolled pattern cannot be shared or stolen; therefore, positive verification or identification can be assured within the limits of the access control system's capability. In a verification scheme, the person attempting entry makes a claim as to his or her identity. This can be a reference template on a smart card carried by the individual, a PIN number, or any other method of electronic identification that references a template stored on the system. The biometric system will compare the reference template associated with the claimed identity to the reading from the user seeking access to see if they match. This is also known as "one-to-one matching" (represented as 1:1). The system matches a single reference template associated with the claimed identity to a reading resulting from an attempted access. Identification consists of comparing the mathematically derived template of the measured characteristics from an access request event to a database of enrollment templates to identify the individual. This answers the question, "Who am I?" Identification is also known as a "one-to-many matching" (represented as 1:N). The system searches through a database of some number (N) of templates for a single template that corresponds to the one from the access request. Most databases are indexed in some way to reduce the number of reference templates that actually need to be compared to make the search easier and faster. Biometric access control system sensitivity can be adjusted to favor security or convenience. Since biometric systems are pattern-matching systems, the allowable variation from exact conformance to the enrolled pattern is adjustable.

A biometric device should provide a function that is complementary to other systems and measures of a physical protection program. Critical considerations for biometric devices include the following:

2.1.3.1.1 Accuracy

What are the false accept and false reject rates for the system in the given configuration?

2.1.3.1.2 Throughput Rate

Does the system operate quickly enough to process the traffic entering the facility?

2.1.3.1.3 Environment

Can the measurement be made accurately in the background environment? For example, the system may be unable to acquire clear voice samples in a very noisy area or acquire accurate facial images where sun glare can strike the individual's face during parts of the day.

2.1.3.1.4 Database Access

How quickly can the template be retrieved from a central database? Some adaptive biometric systems adjust the template to account for human changes over time. How will these updates be communicated to the database and distributed to other access points?

2.1.3.1.5 Tamper Resistance

How well are the measuring devices protected from tampering? Are the electronics that perform the match and control door locks protected from tampering? Are all templates, databases, and communication lines protected from tampering?

2.1.3.2 Facial Recognition

Facial recognition is a biometric access control technology that uses one or more photographic images to recognize a person by measuring points on a face under controlled conditions. Facial recognition systems are not intrusive, require no physical contact with the user, and enjoy high user acceptance. These systems can be used for both verification and identification requirements. A number of technologies are used in facial recognition access control systems. The two major categories are video imaging and thermal imaging. Video facial recognition analyzes the unique shape, pattern, and positioning of facial features by mapping those features to create a mathematical model. Video can be further subdivided into two-dimensional (2-D) and three-dimensional (3-D) imaging. The number of cameras used is the primary physical difference between 2-D and 3-D systems. A 3-D system uses two cameras and integrates the images to create a 3-D digital template. In 2-D systems, a single camera acquires the image to produce a 2-D model. Thermal imaging uses an infrared camera to produce a facial thermograph. The system digitizes the thermal pattern resulting from the heat produced by the blood vessels under the skin.

Facial recognition for access control purposes consists of four steps: sample capture, feature extraction, template comparison, and matching. Enrollment is straightforward, requiring 20 to 30 seconds to take several pictures of the enrollee's face. This photography sequence is best done with varying angles and expressions to allow for more accurate matching. The system extracts the relevant information and uses mathematical techniques to create a reference template that is stored in the database. Verification or identification follows the same steps as enrollment. The user "claims" an identity through a login name, smart card, or terminal entry, then sits or stands in front of the camera for a few seconds. The system extracts the template, compares it to the reference template, then grants or denies access. The point at which the two templates are similar enough to match, known as the threshold value, can be adjusted for different persons, time of day, and other factors.

Access control facial recognition should not be confused with law enforcement surveillance systems, which try to match faces from very large populations in public places with a database of persons of interest. This is an entirely different and much more difficult problem than access control. Facial recognition is useful for indoor verification and identification applications in which the ambient lighting and environment can be controlled. It is not designed for outdoor use or in situations where lighting is variable. Facial recognition is not recommended for areas where personal protective equipment involving face masks is required.

2.1.3.3 Fingerprint Recognition

Fingerprint recognition is one of the most widely used biometrics in the access control industry, in part because fingerprints are one of the oldest forms of personal identification. Classification systems still widely used in law enforcement and forensics were developed in the late 19th and early 20th centuries. Biometric fingerprinting for access control purposes does not use these classification systems. Biometric fingerprint access control applications use one or both of two fingerprint characteristics: minute details (minutiae) and ridge patterns. Minutiae recognition is the most common form of biometric access control fingerprint identification.

Minutiae include the discontinuities that interrupt the otherwise smooth flow of fingerprint ridges, and the abrupt ridge endings and bifurcations (the point where one ridge divides into another). The ridge-pattern-based comparison technique is performed in two major blocks: image enhancement and distortion removal. When a finger is applied to a measuring device, the ridge pattern is distorted to a certain degree. The key to accurate comparison of the ridge pattern is the ability to define and remove the distortion.

Several different methods are commercially available for collecting fingerprints electronically. These include optical, silicon chip, and ultrasound technologies. Optical technology is the oldest and most widely used collection method. For optical collection, the finger is placed on a proprietary, coated platen. A charged coupled device converts the image of the fingerprint (dark ridges and light valleys) into a digital signal. Silicon chip technology has gained considerable acceptance since its introduction to the marketplace in the 1990s. In most chip systems, the sensor acts as one plate of a capacitor and the finger acts as the other. The capacitance between the detector and the finger is converted into an 8-bit grayscale digital image. Silicon generally produces better image quality with less surface area than optical technology. Ultrasound uses high-frequency sound waves to measure the impedance of the finger, air, and platen to generate a signal. Sound waves penetrate the dirt, grease, and ink and can obtain usable prints in some situations that would foil optical systems.

Regardless of the technology applied, several steps are required to convert a high-quality captured print into a compact template. Feature extraction is the basis by which fingerprint technology obtains a usable sample. The exact processes used are proprietary to each access control vendor. The user places a finger, or fingers, on the reader. The reader captures the image and then, based on its design (e.g., optical, silicon), uses complex algorithms to characterize the print for comparison to the template in the access control database.

Fingerprint-based access control systems are used for both identification and verification purposes and are acceptable for both indoor and outdoor use as they respond well to a wide range of environmental humidity and temperature conditions. These systems are not recommended for applications where users wear gloves and may not be feasible where working conditions are such that the ridges and minutiae abrade. Such applications might include construction sites or nuclear or chemical surface contamination areas.

2.1.3.4 Hand and Finger Recognition

Hand geometry access control systems have been commonly used in physical protection programs for almost 30 years and are commonly available in two forms. Full-hand geometry systems measure the entire hand, while finger geometry systems measure only the index and middle fingers. Hand and finger geometry biometrics are automated measurements of hand and finger dimensions taken from a 3-D image. A reader or camera captures up to 96 features of the hand, such as the shape, width, length of fingers and knuckles, distance between joints, and the shape and thickness of the palm. This system is similar to some types of facial recognition systems in that it examines the spatial geometry of the hand and fingers. Surface details, such as fingerprints, lines, scars, dirt, and fingernails, are ignored. Hand geometry access control systems are especially useful in outdoor environments because the detectors are resistant to many environmental factors, such as ambient lighting or temperature changes. Dirty or soiled hands do not affect hand geometry access control devices. Hand geometry readers can also function within a temperature range of 30 to 150 degrees Fahrenheit (F).

Capturing a hand or finger geometry sample is straightforward. The enrollee places a hand, palm down, in a detector containing a flat plate with five pins that guide the placement of the fingers. The detector registers the dimensions of the hand and fingers using a light-emitting diode, a camera, and mirrors. Typically, three placements are required to enroll on the unit, and the enrollment template represents the average of the measurements from the three placements. When a person uses the system, it compares the shape of the user's hand to the enrollment template. If the enrollment template and the hand match to within a certain tolerance (threshold), the system produces an output to the access control system; otherwise, the user is rejected. Some systems maintain a record of accesses and rejections for each user. Templates may reside in one of three places: in the detector's internal memory, in a central database indexed by user, or on portable media, such as a smart card chip.

2.1.3.5 Vein Geometry

Vein geometry access control systems use the patterns formed by blood veins on certain parts of the body (e.g., the back of a hand, wrist, or face) as a means of controlling access. These systems can be used for identification and verification. Vein geometry is a relatively new technology within the biometric access control industry. An infrared camera is used as a scanner to detect and record the pattern of blood veins located directly under the skin. These patterns are unique to an individual and do not vary over a person's life. During enrollment, a user positions the required skin area over the scanner opening while the scanner makes several infrared pictures. The system produces grayscale images of the infrared pictures, which are converted to binary representations. Algorithms extract the useful features from the binary pictures and render them as a mathematical pattern (template) for matching during later validations. To gain access, the user places the required part of the body on the scanner. The system computes a template and compares it to the reference templates generated during enrollment. The system grants access if the template meets the criteria of a match. Vein geometry systems can be used in high-security access control situations requiring identification or verification. They can be programmed to read body areas other than hands when many employees' hands might be otherwise occupied. The nonintrusive nature of the scan and the ease and speed of application enhance the user acceptance rate.

2.1.3.6 Iris Recognition

Iris recognition and retinal scanning are two biometrics associated with the human eye that are often confused by the public. These are two completely different biometric technologies. In retinal scanning, a visible light illuminates the retina, the back of the eye, from a close range. The iris recognition system takes an infrared picture of the iris, the colored part of the eye, from a distance of 3 to 18 inches (7.6 to 45.7cm).

The iris is the plainly visible, colored ring that surrounds the pupil. It is a muscular structure that controls the amount of light entering the eye. The iris is composed of intricate details, called striations, pits, and furrows, that can be measured. No two irises are alike; even an individual person's two irises are completely different. The quantity of unique information that can be measured in a single iris is much greater than that obtained from fingerprints. Iris recognition technology has traditionally been implemented in high-security situations when imaging can be done at a distance of less than 3 feet and a need exists to search very large databases without incurring false matches. Iris recognition is widely regarded as the most accurate biometric methodology because of the rich level of detail that can be gathered. Available systems capture over 240 unique characteristics in formulating the template (over 10 times as much as some other biometric systems). The probability of two irises having the same pattern (even right and left eyes on the same person) is essentially zero. Because the system uses infrared images, only a live iris can register. The distinctive iris pattern is not susceptible to theft, loss, or duplication. The illegal use of an iris pattern from an authorized user is unlikely, and models of the iris that can fool the detector are almost impossible to produce. At death, iris tissue is one of the first body tissues to deteriorate; forensic pathologists use the iris as an accurate estimator of the time of death. This prevents spoofing the system using an actual eye removed from an authorized user.

2.1.3.7 Retina Scan

The retina is a thin nerve on the back of the eye. It is the part of the eye that senses light and sends impulses through the optic nerve to the brain. It is the rough biological equivalent of the film in a camera. Blood vessels, the parts used for biometric identification, are found in the top layer of the retina and form a pattern that is unique to each individual. Retina scanning devices read through the pupil. This requires the user to position his or her eye to within one-half of an inch of the camera and to hold still while the reading is in progress (10 to 15 seconds). Systems provide a light for the user to focus on while the reading is taking place. The blood vessel patterns are measured at 400 points. This leads to a very high level of accuracy compared to biometrics that collect less information. Retina scan systems are pattern-matching systems and work similarly to most other biometrics. The camera collects the sample, the sample is converted to a binary template, and the template is compared to previously enrolled templates from a database for matching. No way to replicate a retina is known, and a retina from a dead person would degenerate too fast to be useful. Therefore, no extra precautions are necessary to ensure that a retina scan comes from a living human being. Retina scanning is capable of identification and verification tasks, and it can be used for standalone or networked solutions.

2.1.3.8 Voice Recognition and Verification

Voice recognition technology uses the unique aspects of human voice patterns to verify the identity of individuals. The fundamental theory for voice recognition is that every voice is distinct and unique enough to identify the speaker. As a person speaks, the shape of the vocal tract changes. The different shapes that the vocal tract assumes contribute to the uniqueness of the voiceprint. Although not suitable for high-security personnel access solutions as a single method of access control, voice recognition technology is being used in lower security applications and in multimodal applications, in combination with other biometrics and other electronic access control technologies. Voice recognition can use any audio capture device. During enrollment, an individual repeats a pass phrase or a sequence of numbers. Pass phrases should be approximately 1 to 1.5 seconds in length. The individual repeats the phrase a set number of times, and enrollment usually is completed in approximately 30 seconds, which is slightly longer than most other biometric access control systems. Electronic voice templates produce among the largest of all biometric files, each occupying up to 10 kilobytes of computer memory in some products. Voice recognition systems, like all biometric access control systems, capture, extract, compare, and match. Once the software captures the analog voice sample, it converts it to a digital signal, applies an algorithm that extracts the data pattern used to compare to the enrollment data (template), and allows or rejects the user's request for access. Applications of this technology include general access control.

2.1.3.9 Signature Dynamics

Dynamic signature verification is a biometric technology that is used to identify a user from a handwritten signature. One of the uses for signature dynamics is physical access control. Signature dynamics technology uses the distinctive aspects of the signature to verify the identity of the individual. The technology examines the behavioral components of the signature, such as stroke order, speed, acceleration, and pen pressure, to construct a multidimensional digital image of the signature. A signature is recorded on the screen of a digital tablet with a stylus. The surface of the tablet screen is pressure sensitive to record such parameters as speed, pressure, and stroke. The stylus on some tablets has a pressure sensitive tip that records writing pressure in addition to the measurements from the screen.

During enrollment, the user provides several signatures that are similar enough that the system can locate a large percentage of the common characteristics among the signatures. The system develops a template based on the most common and notable characteristics of the user's style of signature. Some systems generate a single reference template, while others record all of the enrollment signatures to be reference templates. These reference templates are either saved in a central database or recorded on a "smart card" that the user carries. The user must be relatively consistent during subsequent signature events for the system to verify identity.

Most signature-based systems modify the reference template with each event to accommodate changes in personal style over time. Signature dynamics access control systems can be used for either identification or verification. These systems are often implemented in controlled access situations where signature or written input processes are already in place (e.g., applications in which written access logs are maintained).

2.1.3.10 Multimodal

Single biometric access control systems often have to contend with unacceptable error rates. For example, approximately 5 percent of the population does not have "legible" fingerprints. Multimodal access is the ability to combine two or more mode channels or biometric traits in the same interaction or session. The most obvious reason for using multimodal biometrics instead of a single system is to make access control operations more secure. The current challenge is to design a biometric access control system with as low an error rate as possible that will cover the entire group of individual users and that cannot be compromised under any reasonable scenario. True multimodality implies the ability to analyze several biometric traits simultaneously, instead of sequentially, which means analyzing different biometric traits one at a time. Sequential analysis takes more time and results in lower accuracy. In some multimodal biometric access control systems, the detector output is sent to the central middleware system. In others, the detector output is sent to a comparator and then, once the match is made, the output is sent to a central middleware system for the final access decision. Essentially, multimodal biometric systems remain pattern-matching applications, with each pattern matched to a reference template and a final score generated. This final score is compared to the template to generate the signal to grant or deny access.

Multimodal biometric access control systems are most feasible for very high-security situations. These systems may use characteristics from the same body parts, or from different ones, but samples are all collected simultaneously. For example, one well-known scheme uses face recognition, voice recognition, and lip movement, and then integrates all the signals at once to produce an access control signal. Another uses hand geometry combined with fingerprint recognition. Recently developed, advanced middleware software can combine any combination of biometric devices, regardless of type or manufacturer.

2.2 Keys, Locks, and Combinations

Locks are devices that can be used to assist in controlling access to areas, facilities, and materials through doors, gates, container lids, and similar material or personnel access points.

The key physical attributes of components, such as doors, gates, and container lids that enable them to function as a physical barrier are the fastening mechanisms (e.g., locks, hasps, hinges) which delay, prevent, and control access. Their effectiveness, however, lies in their use in conjunction with other security measures, such as intrusion detection devices and seals. Although some locks are difficult to pick or manipulate, no lock can claim to be "manipulation proof." Because of the large variety of locks available, it is necessary to subdivide the locks into the following types: (1) combination, (2) key, (3) electrical, and (4) pushbutton mechanical locks.

2.2.1 Combination Locks

Combination locks are used on security containers, vaults, and vault-type rooms to control access to these areas and SNM. The two basic types of combination locks are described below.

2.2.1.1 Mechanical

A mechanical combination lock requires aligning one or more movable numbered combination wheels by dialing a correct combination. When the combination wheels are properly aligned, the locking components are released to permit the locking bolt to retract. Mechanical combination locks are available in hand-change and key-change versions. Changing the combination on a hand-change lock requires removing the wheels from the lock, then changing each wheel to the new number, and replacing it into the case. Changing the combination of a key-changeable lock requires using a special change key that is inserted into the rear of the lock case and turned, thereby unlocking the wheels. This allows the wheels to be changed by dialing in the new combination. After the new combination has been dialed in, the change key is again turned and then removed, thereby locking in the new combination.

2.2.1.2 Electromechanical

An electromechanical combination lock may use a microprocessor to process and coordinate the dialed combination with the lock's correct combination. Federal Specification FF-L-2740, "Locks, Combination," lists the standards for General Services Administration (GSA) security container locks and GSA-approved vault door locks used for the protection of classified material. Three electromechanical locks have been approved under this standard. The model X-07 lock was approved in November 1991, the model X-08 was approved in March 1999, and the model X-09 was approved in June 2002. One combination deadbolt lock device has been approved under Federal Specification FF-L-2890, "Lock Extension (Pedestrian Door, Deadbolt)," for use on pedestrian doors for the protection of classified matter. The model CDX-09 consists of an X-09 combination lock mounted on a deadbolt baseplate. The baseplate is surface mounted to the inside face of the door. The deadbolt baseplate has two essential features. First, it provides a means of latching the bolt in the retracted position, which prevents the bolt from being inadvertently extended. Second, turning a knob on the baseplate will retract the bolt to allow egress. The combination on an electromechanical combination lock is changed by inserting a unique electronic key, having two pins, into the rear of the lock case. This key completes a circuit to allow a new combination to be entered and validated.

2.2.2 Combination Padlocks

Class 1 and 2 combination padlocks that meet Federal Specification FF-P-110, "Padlock, Changeable Combination Lock (Resistant to Opening by Manipulation and Surreptitious Attack)," are resistant to opening by manipulation and surreptitious attack. The Class 1 padlock also provides resistance to radiographic techniques. Combination padlocks are not vulnerable to the usual rapping techniques and are usually resistant to manipulation. However, common combination padlocks made of a cast aluminum alloy are without great strength, have little resistance to forcible attack, and are not weather resistant. Other types of padlocks may be weather resistant, but offer little protection against attack.

2.2.3 Key Locks

As in the case of combination locks, it is desirable for a key lock to be capable of being set for a large number of different keys. A high-quality, six-pin lock with 10 key cutting levels per pin potentially permits 10^6 different keys to be used. However, this large number of key cuts is not as useful as a large number of combinations because less time-consuming techniques for defeating key locks are available. Nevertheless, there is value in specifying at least 10^6 key cuts because it requires careful construction of the lock. It is important that the key cut required to open a lock (bitting of a lock) be changeable to permit changes whenever keys are lost or an employee having access to a key is reassigned to other duties or terminated. Changing the bitting of a lock can be accomplished usually by changing pins, wafers, or levers. To ease the task of a bitting change, some locks have cores that are removable for replacement by means of a special key called a "control key." If all the locks in a given facility are keyed to the same control key, the locks are virtually master keyed because, with the core removed, the problem of opening the lock is elementary. Master keying is undesirable from a security point of view because disassembly and inspection of any lock in the system by a competent person provides access to all the other locks in the master-keyed system. In addition, termination of an employee who had access to a master key would require changing the bitting of all locks set for that master key. To change the bitting of a large number of locks can be costly, but the convenience of master systems has made them a popular choice for the implementation of locking mechanisms to control access. A compromise in this conflict between convenience and security may be to use a nonmastered set of locks for high-security areas and to permit master key sets for other, less sensitive areas.

For effective security, it is necessary for a lock to have some resistance to picking and impressioning (a method used to prepare a key by the impressions of the bitting of a lock on a blank key). In general, this resistance can be provided by precision machining of the mechanisms or by special design features, such as side bars, odd-shaped pins, or a large number of levers. Protection of key locks against forcible attack can be enhanced by the use of hardened steel plates (astrigals or mullions) in front of the pins or side bars. Astragals and mullions are metal or steel strips or pieces used to cover standard door hinges, lock pin areas, or the area where doors used in pairs meet. Astragals and mullions provide an increased level of protection to doors and, when used, should be reinforced and immovable from outside the area being protected. It is essential for the bolt of a lock to be retained in the locked position (deadbolt). In some locks, the bolt is held in a locked position by a spring only. In the case of padlocks, this permits the use of appropriate rapping or shimming techniques and, in the case of door locks, the opportunity to surreptitiously retract the bolt without the use of force.

When key locks are used in lieu of combination locks on doors or gates to material access areas in protected and vital area perimeters and for access to vital equipment, they should provide a high degree of resistance to opening by force and tamper. Key padlocks used in lieu of combination padlocks to secure external barriers should be of rugged and sturdy construction and designed for outdoor use. Such locks should also meet interim Federal Specification FF-P-001480 (GSA, FSS), "Padlock, Key Operated (Resistant to Opening by Force, Pick, and Bypass Techniques)."

2.2.4 Electric Locks

In the most popular electric locks, a signal generated by magnetized elements in a plastic card, or by sequential activation of buttons, is compared with a stored code to activate an electrically operated door strike. In some cases, the magnetic card and pushbutton systems are used together. Combined card and pushbutton systems provide, in general, higher security than card-only systems. The advantages of the electric lock are isolation of the part containing the code from the exposed part of the lock, versatility of programming, and ease of integration into alarm systems. Magnetic card systems have some of the problems of common key locks because a lost or stolen card can be used by an unauthorized person. However, reproduction of a card is more complicated than reproduction of a metal key. Pushbutton systems require memorization of a few digits, usually four, and more time to operate than the magnetic card system. Although the number of possible combinations usually is smaller than in the combination lock system, quality electric pushbutton systems compensate for this by incorporating devices that prevent trial and error methods of surreptitious attack by activating an alarm after a number of unsuccessful attempts, or by introducing a delay after each unsuccessful attempt, which prevents operation of the lock for a short period of time.

It is desirable for an electric lock to have the capability for an easy change of combinations. The part of the lock where the combination is set and the housing of the card reader (if the contents of the housing can reveal the combination) should be protected against tampering by tamper switches connected to the alarm system. Generally, where electric locks are installed, a mechanical lock is also installed as a bypass. This lock should be of a quality commensurate with the discussion in the Section 2.2.3 of this report.

2.2.5 Pushbutton Mechanical Locks

Pushbutton mechanical locks are a type of combination lock utilizing mechanical, pushbutton-activated linkages that connect a gate with an external knob to permit opening of the lock. In this lock, it is difficult to design in penalties for punching a wrong combination as is done in electric locks. Therefore, it is important to have a large number of possible combinations. Provisions for easy change of combinations are desirable. Some locks permit a new combination to be dialed in utilizing an Allen wrench when the lock is open, a procedure similar to that for some combination locks. Others require the replacement of internal parts to change the combination. The mechanical locks appear to be fairly resistant to concealed attack; however, more information is needed on their resistance to forcible attack.

2.2.6 Passwords

Passwords used by various security systems or components should be protected against disclosure to unauthorized personnel. These passwords should also be changed periodically to provide further protection against compromise. Procedures for the use, control, and changing of passwords related to security systems or components are important to ensure proper password management and protection against compromise.

2.2.6.1 Password Generation/Verification.

If employed, password generation or verification software should ensure that passwords are generated using the security features following list:

- Passwords should contain at least eight nonblank characters.

- Passwords should contain a combination of letters (preferably a mixture of upper and lowercase), numbers, and at least one special character within the first seven positions.

- Passwords should contain nonnumeric characters in the first and last positions. Examples of nonnumeric characters are: ! @ # $ % ^ & * ( ) - + = _ ` ~ < > , . / ? ' ; " : | \} {] [a A b B c C d D e E f F g G.

- Passwords should not contain the user ID.

- Passwords should not contain any common English dictionary word, spelled forward or backwards (except words of three or fewer characters); dictionaries for other languages should also be used if justified by risk and cost-benefit analysis.

- Passwords should not employ common names; that is, the password should be checked against a set of common names to validate that the password does not contain any of the names, spelled forward or backwards (assuming that the name is over three characters).

- Passwords should not contain any commonly used numbers (e.g., the employee serial number, Social Security number, birth date, phone number) associated with the user of the password.

- Passwords should not contain any simple pattern of letters or numbers, such as "xyz123xx."

In cases in which the user selects his or hers own password (regardless of whether said password is verified by password verification software), the user should ensure that the selected password is consistent with the security features listed below:

- Password should contain at least eight nonblank characters, provided such passwords are allowed by the operating system or application.

- Password should contain a combination of letters (preferably a mixture of upper and lowercase), numbers, and at least one special character within the first seven positions, provided such passwords are allowed by the operating system or application. Examples of special characters are: ! @ # $ % ^ & * ( ) - + = _ ` ~ < > , . / ? ' ; " : | \} {] [.

- Password should contain nonnumeric characters in the first and last positions.

- Password should not contain the user ID.

- Password should not include the user's own name or, to the best of his or her knowledge, the names of close friends or relatives, employee serial number, Social Security number, birth date, phone number, or any personal information that the user believes could be readily learned or guessed.

- Password should not, to the best of the user's knowledge, include common words that would be in an English dictionary or from another language with which the user has familiarity.

- Password should not, to the best of the user's knowledge, employ commonly used proper names, including the name of any fictional character or place.

- Password should not contain any simple pattern of letters or numbers, such as "xyz123xx."

- Password employed by the user on his or her unclassified systems should differ from the passwords employed on his or her classified systems.

The following example passwords are based on the word "doghouse."

- Eight character: D1gh#2Se

- Seven character: D1gh#2S

- Six character: D1gh#2

- Five character: D1g#2

Because legacy computer operating systems and applications include multiple password authentication processes, some systems may not be able to comply with the eight-character password standard. Those systems should apply the requirements such that the intent of the requirements is met to the maximum extent technically possible. Individuals who have passwords or PINs, or both, should not do the following:

- Share passwords or PINs except in emergency circumstances or when there is an overriding operational necessity

- Leave clear-text passwords or PINs in a location accessible to others or secured in a location whose protection is less than that required for protecting the information or security assets that can be accessed using the password or PIN

- Enable applications to retain passwords for subsequent reuse consistent with the licensee's CSP

Passwords and PINs should be changed as follows:

- At least every 6 months

- Immediately after sharing

- As soon as possible, but within 1 business day after a password or PIN has been compromised or after one suspects that a password or PIN has been compromised

- Upon direction from management

- Upon unfavorable employee termination

If the capability exists in the information system, application, or resource, the system should be configured to ensure the following:

- Three failed attempts to provide a legitimate password or PIN for an access request should result in an access lockout that may be automatically restored, following a predetermined time period decided by the system manager, or manually restored. Alternative responses (e.g., by increasing the delay between attempts with each failure) to three failures to provide legitimate passwords for an access request (e.g., by increasing the delay between attempts with each failure) are also acceptable, assuming that the approved CSP documents such alternate responses.

- When a password specification does not comply with the guidance addressed herein, and when the failure to comply is verifiable by automated means, then the password specification should be rejected.

- After a password has been in use for 6 months, the individual should be notified that his or her password has expired and must be changed within five access requests or lockout will occur.

- Any password or PIN file or database employed by the information system is protected from access by unauthorized individuals as technically feasible.

2.2.7 Personal Identification Numbers

When PINs are used along with a credential for gaining access to protected areas or other secure areas, the following minimal standards should apply. The PIN should be at least four digits long. Users of the system should be allowed to select their own PINs. The PIN should be kept private. No other system operator, security officer, technician, programmer, database administrator, or other individual should be able to retrieve an individual's PIN. At access control portals, keypads provided for PIN entry should minimize the opportunities for observation to prevent compromise of an individual's PIN through observation of its entry by another person. If an individual forgets a PIN, he or she should be required to reenroll in the access control system and select a new PIN. All individuals should be required to change their PINs annually.

2.2.8 Accountability

To reduce the probability of compromise, strict control and accountability measures should be established and implemented for the control and accountability of keys, locks, combinations, passwords, and related access control devices. To accomplish this objective, access control devices that may have been compromised should be retrieved, changed, rotated, deactivated, or otherwise disabled when a reasonable belief exists that compromise has occurred even though physical evidence has yet to be uncovered. Changes should also be made whenever a person who had access to the areas for which the devices were designed no longer requires unescorted access. Discretion should be exercised regarding the change of lock combinations, cores, access codes, and the like when an individual terminates employment under favorable conditions. Keys, combinations, passwords, and other access control devices should only be issued to individuals who require access to perform official duties and responsibilities.

2.2.8.1 Combination Lock Control

The most important aspect of lock control for combination locks is the protection of the combination. Combinations should be changed by a person authorized access to the contents of the container or by the facility security coordinator or his or her designee in the following circumstances:

- The initial use of an approved container or lock for the protection of classified material

- The termination of employment of any person having knowledge of the combination or when the unescorted access or clearance granted to any such person has been withdrawn, suspended, or revoked

- The compromise or suspected compromise of a container or its combination or discovery of a container left unlocked and unattended

- At other times when considered necessary by the facility security coordinator or facility security management

The combination should only be given to persons who are authorized access to the storage containers. Security containers, vaults, cabinets, and other authorized storage containers should be kept locked when not under the direct supervision of an authorized person entrusted with the contents. Losing the combination of a mechanical or electromechanical combination lock can be an expensive situation. To prevent the loss of combinations, a record of the combination should be stored in another location that is as secure as the place protected by the lock, the record must be marked with the highest classification of material authorized for storage in the container, and superseded combinations must be destroyed.

2.2.8.2 Key Lock Control

The security of an access control system based on key locks depends on complete denial of keys to unauthorized persons. A record of each key, the names of individuals to who keys have been issued, and the date of issue and return should be maintained. A complete inventory of all keys contained in a secure storage container and review of the associated key log should be performed at periodic intervals. The inventory interval of a key storage container should, at a minimum, provide accountability of controlled keys once at the end of each security duty shift or every 12 hours, as applicable. Inventory of lock cylinders and control keys should be performed on a periodic basis, at a minimum semiannually, for locks used in the protection of the facility.

A common weakness in mastered key systems is the lack of accountability of lock cylinders. To correct this situation, it would be necessary to require a control system involving the accountability of every mastered lock cylinder that has the bitting in present use, either for the master, or in the case of removable cores, the control key. Lock cylinders and control keys should be maintained in a secure container with access to the container limited to designated authorized personnel. The utilization of controlled security seals with an associated controlled security seal log can be implemented on security containers having limited use. Use of controlled security seals on the container increases protection and provides flexibility in inventory schedules.

2.2.8.3 Electric Lock Control

The security of an electric lock system depends on strict control of combinations and cards. The magnetic codes in the cards and the combinations should be changed whenever an employee with access is terminated or is reassigned. Strict accountability of combinations and cards is recommended.

2.2.8.4 Cipher Lock Control

Similar to other combination locks, the combinations should be changed when employees with access to the combination terminate their employment or are reassigned.

2.3 Search Procedures and Equipment

2.3.1 Overview

Within an access control system, a formalized process for searching personnel, vehicles, and material should be proceduralized and implemented as needed at each access control portal. When combined with access control equipment and measures, the search process should prevent unauthorized bypass and the introduction of unauthorized items into the protected area. This goal can be accomplished through positioning attendant security personnel within the access control area and search train to enable the constant observation; complete search; and positive control of all persons, vehicles, and material being processed. For example, within a protected area personnel access control portal, at least one security force member should be assigned to observe and control personnel entering and processing through the portal detection devices of the search train while another should be operating the X-Ray detection device or physically searching hand-carried items. This access control portal should also be provided overwatch observation by an individual responsible for final access control to the protected area. The objective of this implementation method is to prevent attendant security personnel from becoming overwhelmed with multiple access control responsibilities and to enable an increased level of attention to the search and control functions which contributes to the effectiveness of the overall access control process. The number of attendant security personnel assigned to the access portal should also account for the throughput rate of the portal in concert with the need to maintain a high level of attention to the search and access control process. The number of attendant security personnel assigned to access control portals should be periodically evaluated to account for changes in site-specific conditions.

Search equipment configuration and layout should provide the attendant security personnel optimum oversight, control, and observation of each person processing through each component of the search process. These areas should also provide some form of protective shelter (e.g., physical barrier, bullet-resisting protected position, etc.) that attendant security personnel can use in the event of an attack. The search area should be spacious enough to accommodate all required search processes with minimal potential to compromise the integrity and sanitization (verified condition of being absent of unauthorized items) of ongoing and completed searches. Each access control portal should include a designated area for physical (e.g., hands-on pat down, portable hand-held detector) searches of personnel, materials and vehicles that warrant an additional search. Considerations for the placement of this area should include its proximity to the facility protected area boundary to preclude unauthorized entry and its proximity to other personnel, materials and vehicles processing through the search train for the safety and integrity of the searches being conducted. Other considerations may include additional search equipment and an alternate access portal that will facilitate the search and processing of larger packages, materials and vehicles.

Search processes for personnel, vehicles, and materials should only be conducted by trained and qualified security personnel using search equipment or hands-on/pat-down searches, or a combination of both, consistent with the physical protection program design goals and facility implementing procedures. When searches are being conducted, some form of overwatch observation should always be provided to support the integrity of the search, control access and initiate a response to a threat if needed. Methods of overwatch observation can include direct (e.g., personnel present), and indirect (e.g., video surveillance) observation or a combination of both and should be consistent with the level of physical controls inherent to the access control point and search location. For example, searches conducted in areas that provide minimal physical controls (e.g., physical barriers) should be provided both direct and indirect observation. In such areas, the direct observation should be provided by an armed overwatch that is present within the immediate vicinity, and capable of providing immediate response while the indirect observation should be provided through video surveillance that is monitored so that an additional response in support of the immediate response can be initiated as necessary. In this example, the armed overwatch should be responsible for ensuring the control of the area designated for search, thus maintaining the integrity of the search and preventing unauthorized entrance beyond established boundaries (controlling access). The individual conducting indirect observation should be responsible for monitoring the access control portal and search area and initiating a response in support of the immediate response initiated by the armed overwatch. The individual conducting indirect observation of the access control portal should also have the capability to initiate access control measures at the portal such as controlling active physical barriers.

Searches should be conducted and completed before allowing personnel, packages and materials, or vehicles to proceed to the access portal. A search is normally characterized as being complete once the search personnel have determined that the persons, material, packages, or vehicles being searched are without unauthorized items, have met the criteria for entry, and are no longer detained for the conduct of the search process.

To facilitate vehicle identification and search functions, the use of vehicle sally ports (controlled vehicle access passageways) provide an effective means of isolating, identifying, and searching vehicles in a controlled area before allowing entry. When processing vehicles for access to the protected area, vehicles and their operators should be separated for the access control and search processes once the vehicle is in the sally port. Vehicle operators should be processed through the access control and search processes at the personnel access control portal while the vehicle is processed for access and searched in the vehicle sally port.

Vehicle sally port design should include consideration of appropriate crash-test-rated active vehicle barriers that are capable of independent and simultaneous operation. The local operational controls for these barriers should be isolated within a controlled or protected area that is not accessible from outside the sally port and should be configured to provide audible and visual annunciation within the facility's alarm stations indicating the operational status of the barrier. These active vehicle barriers should also have operational controls located within the facility's alarm stations that are capable of overriding the local operational controls for threat- and nonthreat-related emergency situations. Active vehicle barriers should be equipped with an emergency power source (e.g. uninterruptible power supply) or have the capability to be placed in the denial position in the event of a loss of normal power. To prevent tampering or unauthorized manipulation, the motive and power controls along with associated cables and emergency power sources for active vehicle barriers should not be accessible from outside of a controlled or protected area. These system components should also be provided a level of protection (e.g., bullet-resistant hardening, conduit) that prevents or delays the function of the barrier from being affected by external attack before the barrier can be placed in the denial position. A system possessing these features provides multiple options for performing vehicle access control functions and affords an increased level of security.

Other considerations include equipping the vehicle sally port with search equipment designed for use on vehicles or portable search equipment, such as portable explosive trace detection devices. These devices provide additional search capabilities when a complete physical search of vehicle cargo is not practicable or is deemed hazardous.

Consideration should be given to equipping access control portals with one or more duress alarms placed in concealed locations that are accessible to attendant security personnel and can be activated in an unobtrusive manner. An acceptable alternative to permanently installed duress alarms is the use of portable duress alarms that can be worn or carried. Portable, two-way radio communication devices intended for security and law enforcement personnel are commonly equipped with duress signaling capabilities and may also be used to indicate duress.

Attendant security personnel that conduct searches should not rely solely on the alarm annunciation and alarm indication of search equipment to assess personnel, vehicles, and materials for unauthorized items before granting access. Attendant personnel should use basic observation skills and techniques to determine when additional physical searches are warranted. Examples that may warrant an additional physical search may include personnel performing unusual actions; unusual physical characteristics of personnel, vehicles, and materials that are indicative of concealment; and items that cannot be positively identified. The use of physical searches in conjunction with equipment searches increases the ability to mitigate the potential of unauthorized items successfully passing through the search area.

2.3.2 Physical Search

A systematic approach should be considered when performing physical search functions. Process elements, such as selecting a common starting point, conducting the search by areas or priorities, and transitioning from area to area in a systematic manner, should be applied as best practices to ensure thoroughness. The design and development of search procedures and training should consider such practices.

Procedures for conducting physical searches should consider the safety of personnel assigned to conduct the search. These procedures should ensure that security personnel performing search functions are protected from items such as sharp or pointed objects, which may be secreted in clothing, packages, and containers or on vehicles. Personnel conducting these searches should wear personal protective equipment capable of preventing injury from these types of objects.

2.3.2.1 Personnel

The physical search of personnel (as when using electronic portal search devices) should include the removal of personal property and outer garments that could conceal unauthorized items. All items that are removed should satisfy the established search criteria before being returned to the owner. The following is an example of a systematic process when conducting a personnel search:

(1)   Instruct the individual to remove all items from pockets and search all items removed.

(2)   Instruct the individual to remove shoes, headwear, and outer garments (coats, sweaters), and search all items removed, including both interior and exterior pockets, seams, and linings.

(3)   Search head area.

(4)   Move to collar area, checking inside and under the collar.

(5)   Move to front shoulder area, checking from upper arms to hands.

(6)   Move to front upper torso, checking from upper front torso to waist.

(7)   Check sides of torso, sweeping from underarms to waist.

(8)   Move to waist, checking inside waistband by sweeping from front to sides.

(9)   Move to front crotch area and upper legs, sweeping from crotch area and upper legs to feet.

(10) Move to back upper torso, checking from upper back torso to waist.

(11) Move to waist, checking inside waistband sweeping from back to sides.

(12) Move to buttocks and back upper legs, sweeping from buttocks and back upper legs to feet.

(13) Instruct the individual to lift feet, and check under feet.

2.3.2.2 Vehicles

Before conducting a vehicle search, the vehicle operator should be removed from the vehicle and sent to a location distant enough from the vehicle that the operator cannot compromise the vehicle search process. The operator should be searched before being allowed back into the sanitized vehicle. Where vehicle sally ports are used, once the vehicle is in the vehicle sally port, vehicle operators should be required to dismount the vehicle and proceed to the personnel access control portal to satisfy access control and search processes. When vehicles are entering the vehicle sally port, the inner most active vehicle barrier should remain in the denial position. Once the vehicle operator has left the vehicle sally port, the outer barrier of the vehicle sally port should be secured to establish a controlled environment in which to conduct the search. The following is an example of a systematic process when conducting a vehicle search:

(1) Search the undercarriage starting from front to back, driver to passenger side (inside and around frame rails and bumpers, under and around the engine, wheels, wheel wells, around transmission, above and around any tanks).

(2) Search engine compartment starting from the front to back, left side to right side, top to bottom (in and around all components, opening and searching all compartments (e.g., air filter housings)) that can be opened.

(3) Search the vehicle cargo area starting from front to rear, top to bottom, driver to passenger side (if the area contains cargo, the cargo should be broken down and searched to the lowest possible level without making the contents unusable so that it can be positively identified).

(4) Search the vehicle cab starting from front to back, top to bottom, driver to passenger side (headliner, sun visors, top of and under dashboard, under and behind seats, under seat pads), and opening all compartments (e.g., door compartments, consoles).

2.3.2.3 Material

For safety purposes and to ensure that the integrity of the material remains intact, initial identification of material should be made through observation, checking shipping documentation or product labels, checking a bill of lading, or similar actions. Once the initial identification is made, the material should be removed from their packages or containers, down to the lowest level without making the contents unusable. This process ensures that the material is positively identified and verifies that the material, along with the packaging or container, does not contain unauthorized items. The following steps should be used to search material (in this case, a pallet of shrink-rapped boxes destined for the protected area) to the lowest level without making the contents unusable:

(1) Move the pallet to a controlled location.

(2) Check the shipping documentation. (Documentation identifies 25 boxes of individually wrapped pairs of sterile gloves, 100 pairs per box.)

(3) Remove the shrink wrap and boxes from the pallet.

(4) Open each box one at a time, removing each packet containing a pair of sterile gloves. Visually and physically inspect each packet without degrading the seal of the sterile packaging. Verify that each packet contains a pair of gloves.

(5) Visually and physically inspect empty boxes.

(6) Visually and physically inspect the pallet.

(7) Replace the glove packets in the boxes and then replace boxes on the pallet.

2.3.2.4 Guidance for Personnel Conducting Searches

Personnel conducting searches should be aware that each situation can present its own unique circumstances. In all cases, common sense and a questioning attitude should be the presiding factor. Personnel should not conduct operations or actions that would place personnel, equipment, or the facility under unnecessary risk. When attendant security personnel have cause to suspect the attempted introduction of unauthorized items, the attendant personnel should do the following:

- Deny access to the suspect person, vehicle, or material.

- Call for assistance or notify appropriate security personnel to initiate response.

- Conduct a physical search to confirm the presence or absence of unauthorized items.

- Control the suspect person, vehicle, or material.

- Notify security supervision and the local law enforcement agency (LLEA) as necessary.

Unauthorized items discovered during the search process should be controlled and a chain of custody should be established and maintained throughout the control process including and up to the transfer of the item to the LLEA, where appropriate. Notification procedures, processes for the control of unauthorized items, and level of assistance to be provided by the LLEA should be determined through liaison with LLEA representatives.

2.3.3 Metal Detection

Metal detection is widely used for the detection of metallic components in weapons such as guns and knives. It can also be helpful for the detection of wiring, batteries, and other metallic components of bombs and incendiary devices, as well as for the detection of metals used to shield SNM. While multiple interactions occur between magnetic fields and metal, the primary effect used to detect metal in portal metal detectors is the induction of eddy currents.

Most walkthrough portal metal detectors are weather resistant, constructed with plastic or metal, and possess modular electronics for serviceability with secured and tamper-proof controls. This type of detector provides various levels of detection (i.e., head, shoulders, waist, and ankle). Microprocessors are used in both the detection and control circuitry. Portal detectors provide uniform coverage from head to toe for precise target evaluation, and few adjustments are required to set up and operate the detector.

Hand-held, wand-type metal detectors may be used when the portal metal detectors are inoperable. Most hand-held, wand-type metal detectors offer the following features: (1) low battery consumption, (2) high-speed alarm and reset capabilities, and (3) improved sensitivity and precision accuracy in locating even extremely small masses of all metals. Most brands of hand-held, wand-type metal detectors have been medically tested and approved with no risk to persons with pacemakers and noninterference to magnetic recorded material.

Hand-held, pocket-sized metal detectors are lightweight (6.4 ounces (18 grams)) and compact (11.5 centimeters (4.5 in.) x 6.5 centimeters (2.6 in.) x 3 centimeters (1.2 in.)). The compact hand-held metal detector is simple to use, enables the operator to locate small or large metal objects, and is suitable for a multitude of applications for general security and law enforcement work. This type of hand-held metal detector gives the operator a very positive and silent vibration alarm in the palm of the hand when a metal object has been detected.

Hand-held and portal metal detectors should be capable of detecting prohibited weapons with a highly accurate effective detection rate, and minimal false alarm rate when the detector sensitivity is adjusted to the desired detection level. The devices should be adjusted to discriminate between typical firearm and nonfirearm masses of metal. Annunciation of metal and explosives detection equipment should be both audible and visual.

Glove-type metal detectors, unlike the wand or pocket-sized metal detectors that are held by the operator, are worn on one or both hands of the operator. This type of metal detector was specifically designed for pat-down and hand-search procedures used when the operator wants to be in close contact with the subject being searched. When the operator has located a metal target, the detector gives an alarm indication by means of a vibration against the wearer's wrist. As the vibration is to the operator's wrist and not to the hand, the person being searched does not feel vibration. This offers the operator a covert approach to pat-down procedures. This type of detector also gives the operator's fingers complete dexterity to feel simultaneously for concealed objects that are of a nonmetallic nature. This type of detector is fully automatic and requires no adjustments for use. There are no visible wires, switches, or buttons. Everything is housed in a machine washable, soft stretching neoprene material (such as used in wet suits) or Kevlar for added protection during pat-down procedures. The operator just puts the glove-type detector in the "on" mode and begins a pat-down search.

2.3.3.1 Effects of Eddy Currents

When an electrically conductive material is exposed to a time-varying magnetic field, a voltage develops around any conductive path within the material. This induced voltage causes a current to flow within the material. This induced current is called an "eddy current" (Figure 1). The eddy current in turn produces its own magnetic field, which the metal detector detects.

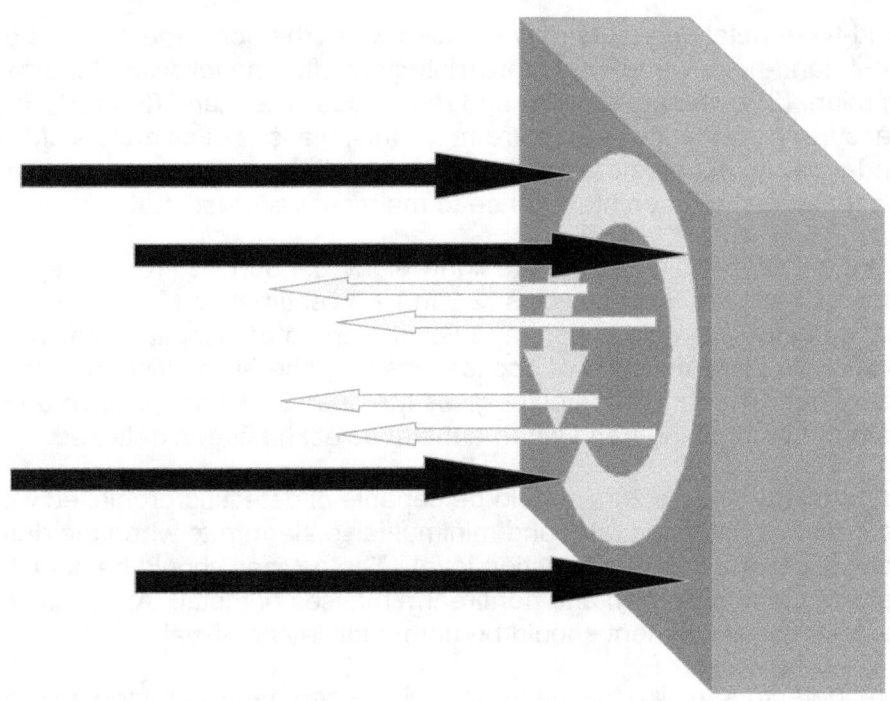

Figure 1: The induced eddy current and the resulting opposing magnetic field

The eddy currents induced in the target metal and their associated magnetic fields die out very quickly in small, highly resistive metal objects, while those in large, highly conductive objects persist somewhat longer. Larger voltages tend to be induced in larger objects and in more highly conductive metals. Another factor affecting the magnitude of the induced voltage is the ferromagnetic properties of the metal. Metals with strong magnetic properties (a large relative magnetic permeability) will have a larger induced voltage than an object of the same size that has a low relative magnetic permeability.

In practical terms, modern portal metal detectors consist of transmitter coils and receiver coils. The transmitter coils generate a magnetic field within the portal's detection volume to induce eddy currents within any metal being carried through the detector. The receiver coils are used to detect the magnetic field produced by the eddy currents.

2.3.3.2 Pulse Detection

Most modern portal metal detectors use a technique called pulse detection. The pulse detection process is similar to when a loud burst of sound (such as a hand clap) occurs near an acoustic musical string instrument. The strings of the musical instrument will respond to the sound by vibrating for a brief period of time. A person could detect a musical string instrument in a dark room by clapping and listening for the response. For the metal detector, the clap is replaced by magnetic pulses, and the response is magnetic energy rather than sound energy. In the intervals between the pulses, the detector "listens" for the return signal from any metal within the detection volume. The detection circuit can perform some signal processing on the received signal, which can help distinguish between signals from metal objects passing through the portal and signals from other moving metallic objects in the general vicinity.

Metal detectors of the pulsed-field variety generate electromagnetic pulses in the range of 60 to 1,500 pulses per second. The shapes of the pulses vary, but they generally have complex shapes; impulses, step functions, and square waves are common. The frequency spectrum of the waveforms contains frequency components from direct current to 20 kilohertz. Coil configurations and signal processing vary from one manufacturer to the next, but the basic technology is the same. An electromagnetic pulse on the order of 1 to 3 gauss is transmitted by means of a coil built into the archway. After an interval following the transmitted pulse, the input from the receiver coil is sampled during a time window. The length of the interval after the transmitted pulse and the duration of the receive window depend on the operating mode (program) and the specific type of metal to be detected.

2.3.3.3 Amplitude Detection

One way that a metal detector processes signals is simple amplitude detection. Only signals of sufficient magnitude will cause an alarm. The user can adjust the threshold at which the signal causes an alarm. Some detectors can also analyze the persistence of the return signal (how long the rapidly decaying signal lasts) and detect the phase of the response to determine the type of material present. For instance, a detector may be adjustable to detect ferromagnetic materials (such as steel) strongly while weakly detecting nonferromagnetic materials (such as copper or gold). In addition, by using multiple transmitter and receiver coils, some detectors can determine whether the return signal is being generated by the presence of a single large object or several smaller objects. Some detectors that use multiple transmitters and receivers can also determine and indicate on a display the approximate location of the source of the alarm. This last feature may greatly reduce the time required to resolve portal alarms using hand-held metal detectors.

2.3.3.4 Factors Affecting Metal Detection

Many factors in a metal detector's environment can have a strong influence on the detector's performance. Any large metallic object located near a detector can distort the transmitted magnetic field and potentially create regions of increased or decreased sensitivity. Specifically, a metal detector in a portal cannot restrict detection to between the portal panels; the detection extends to approximately 2 meters beyond the portal panels. Outside the panels, the detection is weak, and small items will not likely cause alarms unless they are very near the outside of the transmitter and receiver panels. However, large items (such as metal doors) produce signals large enough to cause alarms. In one documented case, a metal detector was subject to detection of large vehicles passing through a vehicle portal several meters outside the pedestrian portal where the detector was installed. Many sources of electromagnetic noise can interfere with metal detector operation, as described below:

- Nearby metal doors can cause false alarms.

- Metal cabinets can cause false alarms when drawers open and close and can also cause problems with the sensitivity profile.

- Installation concerns include stability of floors. People moving near the detector can cause floor vibrations that induce motion in the detector, causing alarms. The detector should be anchored to the floor or some other fixed structure to prevent movement.

- Metal pipes can move when liquids move through them and cause alarms.

2.3.3.5 Continuous-Wave Techniques and Hand-Held Metal Detectors

Hand-held metal detectors can use pulse or continuous-wave techniques for detecting metal. Hand-held detectors can be used as the primary tool for screening people but, because of the time required to thoroughly screen an individual, the portal metal detector is preferred for primary screening. A hand-held detector can then be used to resolve alarms that occur in the portal. Since a hand-held detector senses metal at a short distance, only a small part of the human body is scanned at a time. Operators need a well-defined and well-designed screening procedure for a hand-held detector. An appropriate procedure is described below.

The use of a hand-held metal detector alone is not always sufficient to identify the source of an alarm. For example, a hand-held metal detector will alarm on a belt buckle; it will not indicate whether a metallic item is behind the belt buckle. Asking the individual to remove the buckle and then screening the beltline with the hand-held detector again or performing a pat-down inspection of the beltline area is required to establish that the belt buckle is the only source of the alarm. A variable pitch or volume hand-held detector can provide sufficient information to make this judgment.

A hand-held metal detector allows items to be located quickly, minimizing the extent of hands-on search in personnel screening. The loop at the end of the detector wand should be held parallel to (not perpendicular to) the body being screened. The hand-held detector should be swept over the body at a distance consistent with manufacturers recommendations (usually 1 to 4 inches) in a pattern similar to the following:

(1) Starting from the shoulder, sweep down the front of the body to the foot region (including bottom of foot), then to the other foot (including bottom of foot) and back up the opposite side of the body, ending with the opposite shoulder. If the detector's scanning coil diameter (or length) is less than half a person's body width, modify the pattern to ensure adequate coverage.

(2) Repeat the same pattern over the back side of the body.

(3) Starting at one shoulder, sweep the detector coil over the outside of the arm from the shoulder to the bottom of the sleeve, then up the inside of the arm to the armpit. Sweep down the side of the body to the foot, then up the inside of the leg and down the opposite leg and back up the other side of this leg. Repeat the sweep of the inside and outside of the arm opposite the side where the sweep began, ending at the shoulder.

(4) Sweep the head area after asking the person to remove all headgear.

Particular attention should be paid to the pocket areas. Variations of the above pattern are acceptable. However, care should be taken to ensure that the entire body is covered, and it is recommended that each search follow the same pattern to prevent incomplete searches.

Hand-held metal detectors are sensitive and will likely sound an alarm when passed over very small metallic objects, such as rivets on pants, metal buttons, and brassiere underwires. For this reason, a detector that has a variable intensity or pitched tone that provides some indication of the size of the metallic item being sensed is useful. With a skilled operator, a detector of this type allows the entire search to be performed with no physical contact with the person being scanned. If, on the other hand, the detector simply sounds a constant tone when sensing metal of any size, a pat-down of the area may be required to ensure that the item being sensed is not a weapon.

Each search should be a complete body search. Stopping the search after finding the probable cause of the portal detector alarm does not ensure that there are no other items on that person. In addition, some medical surgical implants, such as knee and hip replacements, can cause portal metal detector alarms. The use of a hand-held metal detector and a pat down of the area can verify the cause of the alarm. However, it is again important to continue the search to cover the entire body.

2.3.3.6 Availability

Active metal detectors are available with a variety of prices and performance levels. Low-cost systems are capable of easily detecting normal-sized handguns (such as a compact .25-caliber automatic or larger), but may have difficulty detecting smaller items, such as small knives and highly compact guns constructed of more exotic materials (e.g., composites, stainless steels, and other alloys). Certain highly desirable features, such as alarm location displays, will also increase cost.

2.3.3.7 Magnetometers

One often hears someone speaking on the news about "magnetometers" being used in security systems at airports or elsewhere. More likely than not, that person is incorrectly referring to metal detectors as magnetometers. A magnetometer is a very specific passive device that monitors the earth's magnetic field and senses disturbances in that field. The only materials that have a profound local influence on the ambient magnetic field are ferromagnetic materials. Ferromagnetic materials are those materials that are strongly attracted to a magnet. The materials that exhibit ferromagnetism are iron and nickel and certain alloys of those elements. Magnetometers do not detect materials such as copper, aluminum, and zinc. The reason that magnetometers have been used for contraband detection is that most guns manufactured today are composed of steel, an alloy of iron.

2.3.4 X-Ray Imaging

In general, the use of X-Ray-based search systems is intended to assist access control personnel in determining that packages and other hand-carried items entering a secure area do not introduce unauthorized items into the area. One of the most visible applications of this technology is in airports. People who travel by air are accustomed to seeing their carry-on baggage screened by X-Ray-based systems where humans view the X-Ray image in an attempt to identify unauthorized items. However, modern search systems have enhanced capabilities in analyzing the imaging signal to help the screener identify such unauthorized items. Systems with enhanced capabilities use analysis of transmitted or backscattered X-Rays, or both, to assist the operator by providing additional information on the nature of the materials inside the package.

2.3.4.1 Z Number

Materials can be characterized in many different ways, but for X-Ray imaging one of the most useful ways is to distinguish between high-Z and low-Z materials. The Z number of an element is its atomic number. For example, the atomic number of iron is 26. Heterogeneous materials such as compounds are not composed of a single element, so the compound will have what is called an effective Z number. This effective Z number is determined by the Z numbers of the various elements and the fractional component of each within the compound. From an X-Ray absorption standpoint, a somewhat arbitrary division is often used where Z numbers below 26 are categorized as low-Z while 26 and above are categorized as high-Z numbers. Thus, organic materials are low-Z materials and most metals are high-Z materials. The Z number of materials (both Z number and effective Z number) determines how strongly X-Rays are absorbed. The higher the Z number, the more the material absorbs X-Rays.

In general, low-Z materials absorb X-Rays weakly and will appear translucent (i.e., will be difficult to see) in an X-Ray transmission image, while high-Z materials will appear dark. Weapons are typically composed of high-Z materials and will appear dark in the image. Weapons tend to have distinctive shapes, so an operator's ability to recognize shape is very important in detecting these threat objects. Low-Z threat materials, such as explosives and drugs, lack distinctive shapes so they may be difficult for an operator to detect without additional assistance from the imaging system.

2.3.4.2 Dual Energy and Backscatter Imaging

Two technologies that are helpful in assisting an operator in spotting threat materials are dual energy and backscatter imaging. These technologies were developed to enhance the image to assist an operator in detecting threat materials but have also been used to develop automated bulk explosives detection systems.

2.3.4.2.1 Dual Energy Systems

The dual energy technique exposes the package to either of the following:

- Two distinctly different energies of X-Rays

- A range of energies and analyzes the transmitted energy at the different energy levels

By examining the differences between the interactions with X-Rays of different energy, some discrimination between various types of materials can be made. Simple black-and-white transmission systems cannot easily distinguish between thick sections of organic low-Z materials and thin sections of high-Z materials. Even more difficult is distinguishing between benign low-Z materials and threat materials, such as explosives. Dual energy systems can distinguish between these materials and highlight the difference using color (e.g., low-Z materials are displayed as orange, while high-Z materials are displayed as green).

2.3.4.2.2 Backscatter Systems

All materials will backscatter some of the X-Ray energy when exposed to X-Rays (i.e., X-Ray energy is sent back toward the X-Ray source). Low-Z materials are much more efficient at backscattering than high-Z materials. When backscatter energy is captured and used to produce an image, the low-Z materials will appear bright against a dark background. For this reason, backscatter imaging can easily distinguish between low-Z and high-Z materials. A typical backscatter system will have two monitors: one that displays the transmission image while the other displays the backscatter image. These systems expose the package to the smallest radiation dose of all X-Ray-based systems.

2.3.4.3 X-Ray Imagers

X-Ray imagers are manufactured around the world and vary in price, which is usually consistent with the performance characteristics and features desired. Some common features include the following:

- Reverse image (similar to a negative image where some items may be more visible)

- Image zoom

- Edge detection

- Threat image projection (TIP) for training operators

- TIP-enhanced systems contain a library of various threat images that can be projected into the image of a package that does not contain any threat items to help an operator learn what to look for and to test operator alertness and effectiveness.

2.3.5 Computed Tomography

Computed tomography (CT) scanning (sometimes referred to as "computed axial tomography" or "CAT" scan) for the detection of contraband originated in the medical industry. Early CT scanners were modified medical systems that were used for contraband detection, primarily the detection of explosives in checked baggage at airports.

2.3.5.1 Operational Concept

In these systems, an X-Ray source projects a fan of X-Ray energy through a package being scanned onto a curved array of photodiodes (Figure 2). The source and the diode array are mounted on a circular movable carriage. The source and the diode array take readings as they are rotated around the package. The information from the scan is sent to a computer that converts the information into a 2-D slice. The package is moved forward slightly and the process is repeated. When the 2-D slices are stacked, they form a 3-D picture of the package and its contents. Since the CT scanner can calculate the volume of a body internal to the package and the transmission properties of that body are in the scan information, the scanner can determine with fair accuracy that the material does or does not have X-Ray transmission characteristics (effective Z number) typical of explosives.

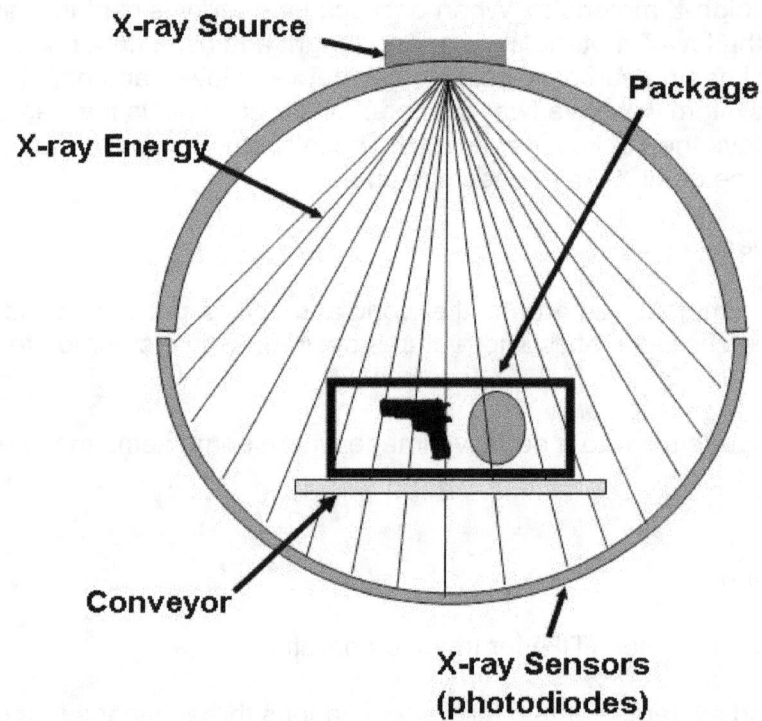

Figure 2: Conceptual design of a CT scanner

2.3.5.2 Characteristics of CT Scanners

Due to their large size, CT scanners require a relatively large space to not only accommodate the scanner but to employ control measures associated with scanning operations. In addition, they can have a negative impact on camera film. Most CT scanners can accommodate packages the size of a large suitcase and possess highly effective detection capabilities for certain types of explosive materials.

2.3.5.3 System Resolution and Accuracy

Figure 3 shows how the multiple-look angles provide information that allows CT image reconstruction. This simplified example shows how a four-look-angle examination of a point in space allows the surface area of a CT slice to be determined with some accuracy. The energy is received by a photodiode with some fixed surface area. The X-Ray energy is typically not composed of discrete beams but rather is a fan of energy. The area of the diode becomes the cross-sectional area of the received X-Ray beam. The smaller the photodiode, the smaller the beam and the higher the resolution of the system. Combining the beam information for all the look angles produces a volume in space called a voxel. Smaller beams result in smaller voxels, which increase the system's resolution. While smaller photodiodes may improve resolution, they are less efficient at collecting energy and may achieve higher resolution at the cost of lower system sensitivity. The smallest photodiode that will produce the desired energy collection becomes the practical limit to the beam-size resolution.

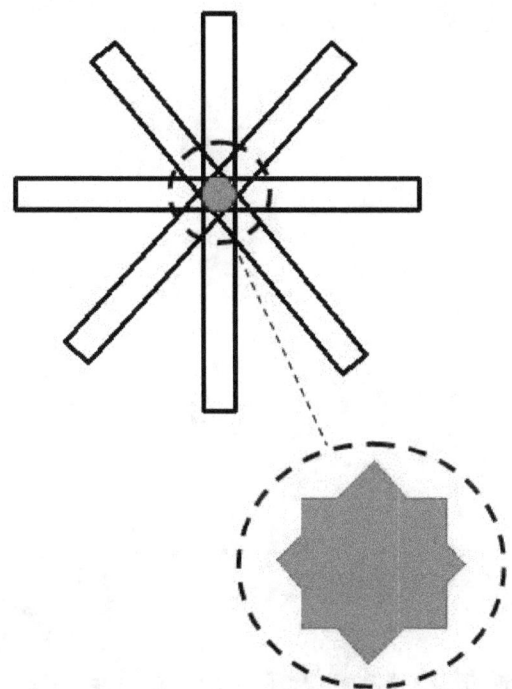

Figure 3: Multiple-look angles

2.3.5.3.1 Number of Look Angles

Another factor in system resolution and accuracy in determining the effective-Z is the number of look angles used in the image reconstruction. In the example shown in the diagram, the four look angles result in "artifacts" (represented by the points around the voxel). These artifacts are regions of uncertainty that result in errors in the calculated voxel volume and the resulting effective-Z. More look angles reduce the size of the artifacts. Actual CT systems use hundreds of look angles.

2.3.5.3.2 Calculating the Effective-Z

Once a series of CT image slices are stacked, the volume of a region in space with the same absorption can be calculated. Since the absorption of that region of space can be measured, the effective-Z of that material can be calculated.

Figure 4 shows an example of the output of a CT detector. The test object is a suitcase containing a sheet of explosive simulant material. The central image represents the region where the potential threat material (explosive simulant) was detected. The displays on the right side are several other CT image slices of the same package. Although it is not clear in this reproduction of the image, the thin line in the middle of the package is colored red to indicate the threat material. This screen image, when displayed, assists in the resolution of alarms by providing the location of the possible threat material. This system is intended to act as an unattended automated alarm system so that an operator is not required to inspect the image for threats.

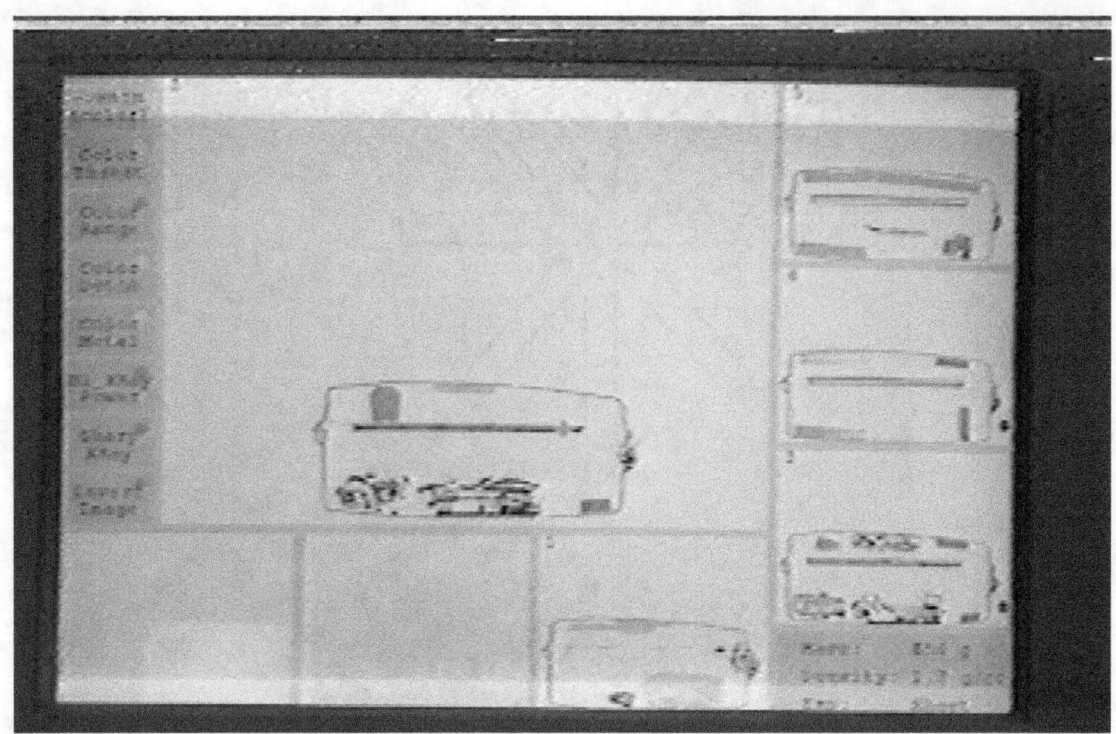

Figure 4: CT scan sample output

Figure 5 is a closeup of the central image of the alarm screen above. The sheet explosive simulant shows more clearly red in this image. The grey object above the sheet of simulant is a votive candle. Candle wax is a low effective-Z material similar in effective Z to polymers. The system has sufficient discrimination to alarm for the simulant but not the nearby candle. Since the material detected is a simulant and not an actual explosive material, it is clear that the system was unable to discriminate between this simulant and an actual explosive. It is recommended that a trace detection technique be employed for verification of detection provided by a CT system.

Figure 5: Enlargement of preceding CT image

2.3.6 Trace and Bulk Explosives Detection

Explosives detectors, whether of the hand-held or portal variety, should be capable of detecting, with at least a 90-percent effective detection rate, explosive compounds in a minimum amount of 200 grams (7 oz.). The false alarm rate should not exceed 1 percent when the detector sensitivity is adjusted to this detection level. Annunciation of metal and explosives detection equipment should be both audible and visual.

Hand-held detectors can detect threats from explosives, chemical agents, toxic chemicals, or narcotics in approximately 15 seconds. This type of detector typically has a cold start time of approximately 10 minutes and weighs approximately 7 pounds (3.2 kg), including the battery and display. Most hand-held detectors are capable of analyzing either trace particle or vapor samples, allowing the operator to apply the ideal sampling technique for the substance suspected. For example, most explosive and narcotic substances do not have a strong vapor presence and in the real world are very difficult to detect by vapor. Therefore, the most reliable collection and analysis method for those substances is particle collection. The nature of chemical agents and industrial toxins make vapor sampling more appropriate for those substances. The ability to analyze either trace particle or vapor samples lets a trained operator decide which sample collection method will yield the most accurate results.

While dual detectors are not hand held, it is typical of commonly used detection devices that employ a hand-held sampling device in conjunction with a related analyzing device to be located in a semipermanent location in close proximity to the inspection site. The dual detector should be located in a relatively protected location to avoid false readings based on outside contamination. The accuracy and reliability of this type of detector is somewhat limited by the thoroughness of the individual collecting the sample. This type of detector has the ability to detect a wide range of substances and is able to adapt as threats and their needs change. By incorporating two detectors in a single unit, the dual detector is capable of detecting and identifying explosives and narcotics during a single analysis, providing the ability to detect a broader range of substances while maintaining a high degree of sensitivity and specificity. The dual detector system is ideal for security applications in that it has the capability to detect a much broader range of explosive substances in addition to types of narcotics.

Portal detectors should be designed specifically to meet the challenge of screening people for trace amounts of explosives or narcotics in a timely fashion and without coming into contact with them. The device must be capable of head-to-toe screening by using puffs of air to dislodge any particles trapped on the body, hair, clothing, and shoes. These particles would then be directed into the instrument for analysis. This technology should be capable of a high throughput, screening up to seven people per minute. Trace amounts of more than 40 substances should be detected and identified in seconds. Results would be displayed in an easy-to-understand fashion.

Drive-through detectors should be designed to address the need for detection of large vehicle bombs with a high vehicle throughput that combines an automated touchless vehicle sampling system with gas chromatograph and ion mobility spectrometer chemical detection. This system is fully automated for continuous operation and is driven by an intuitive icon-based interface that is both user friendly and highly reliable.

Explosive detection is broadly divided into the following two categories of bulk detection and trace detection:

- In bulk detection, a macroscopic mass of explosive material is detected, usually via X-Ray imaging or a probing technique involving incident neutrons or photons that interact with atomic nuclei.

- In trace detection, a microscopic amount of explosive vapor or particle contamination is detected. Trace detection involves the use of either a manmade chemical sensor or an animal that is trained to sniff and detect explosives. The latter is almost always a canine.

2.3.6.1 Ion Mobility Spectrometry

A variety of manmade trace chemical sensors can be used for explosives detection, including IMS, mass spectrometers, electron capture detectors, and others. The following discussion will be limited to IMS. IMS are the most widely deployed trace chemical sensors in explosive detection field applications, with hand-held, benchtop, and personnel portal systems widely available. Several U.S. Government agencies have deployed these devices in the field. An IMS system works in the following manner:

- A sample, which may or may not contain traces of explosives, is drawn into the instrument. The sample may be ambient air (when testing for explosive vapor) or the vaporized components of a sample collection swab that has been used to wipe a particular surface (when testing for particle contamination).

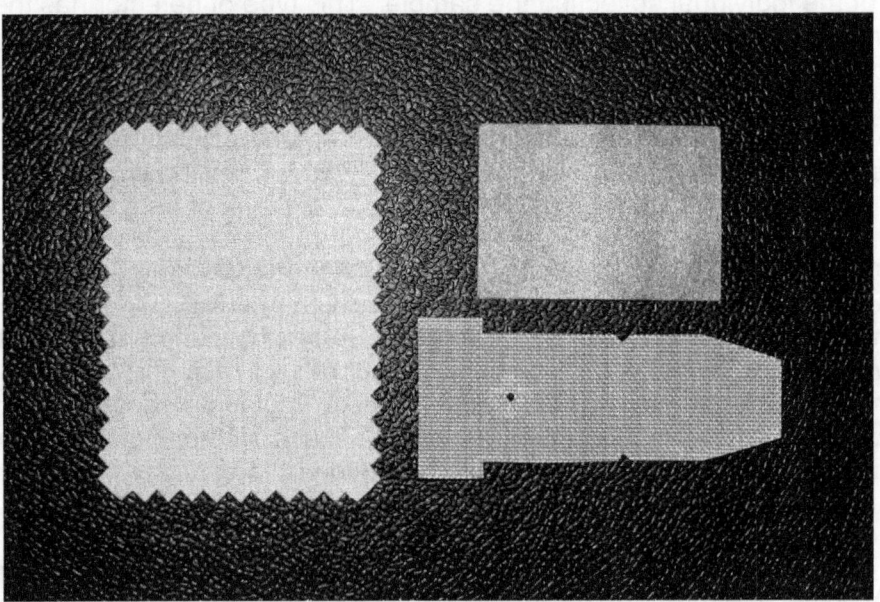

Figure 6: Examples of sample collection swabs used for IMS-based equipment

- The sample enters a part of the instrument called the ionization region, where analyte molecules are ionized, usually by a small, sealed radioactive nickel-63 source.

2-36

- The ions are then periodically pulsed into a second part of the instrument known as the drift region, where the ionized species traverse a certain distance in a certain time under the influence of an applied electric field. This travel time (the drift time) is characteristic of the chemical species involved and can be used to identify particular explosives.

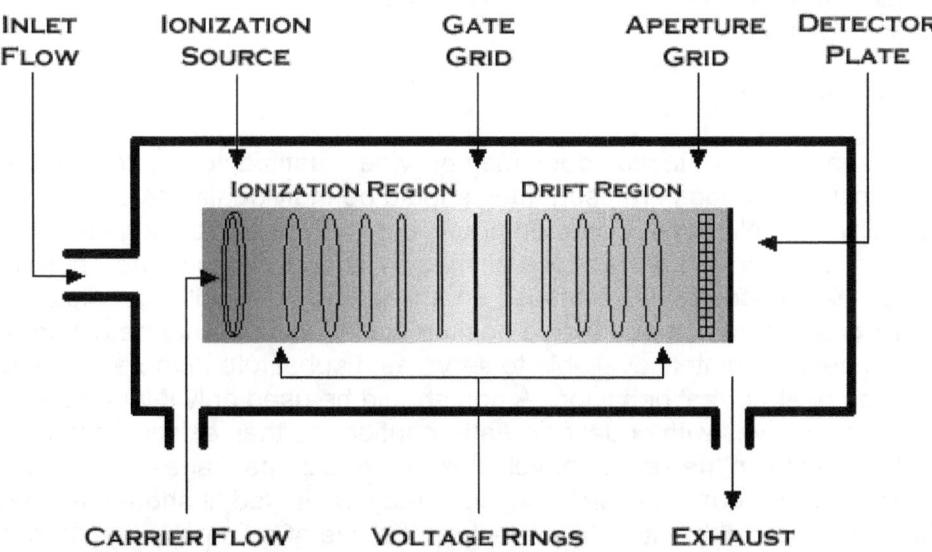

Figure 7: IMS conceptual design

The IMS combines many desirable features, including near-real-time response (a few seconds), subfingerprint sensitivity to a variety of common explosive compounds, moderate cost, and excellent adaptability for field use. The response of an IMS depends to some degree on atmospheric pressure and other environmental conditions, such as humidity. Therefore, recalibration is often required when weather conditions change or when the IMS is moved to a new location at a different elevation.

2.3.6.2 Colorimetric Test Kits

Another means of trace detection of explosives is to employ a colorimetric test kit. These test kits are designed for particle sampling only and consist of sampling swabs and one or more chemical solutions. A swab is wiped on a surface where the user wishes to look for trace explosive contamination. A drop of one of the provided solutions is then applied to the swab, and a color change to a particular color indicates that an explosive of a certain type or category has been detected. If no color change is observed, the initial test is negative, and a drop of another solution is then added to the swab. This process is repeated until all the solutions have been applied to the swab to see whether a color change is produced. In most cases, the solutions must be applied to the swab in a specific order. Characteristics of this type of system include portability, ease of use, low cost and often include the following:

- The test kit has lower sensitivity than the IMS. Limits of detection are typically hundreds of nanograms (one nanogram = 10^{-9} gram) or micrograms (one microgram = 10^{-6} gram) with a colorimetric system versus hundreds of picograms (one picogram = 10^{-12} gram) or low nanograms with IMS.

- In many cases, a color change indicates the presence of one of several different explosives that are chemically related, rather than a unique explosive.

- Maintaining operation requires resupply of perishables (e.g. replacement of the sampling swabs and solutions).

2.3.6.3 Canine Detection

Utilization of K-9 search and detection dogs may provide a satisfactory detection capability for firearms and explosives. If dogs are used, they should be individually tested to ensure their continued capability and reliability. Common practice is to train search dogs and their handlers as a team. Maximum effectiveness is accomplished by ensuring noninterference by other personnel during search operations. Because an animal may present unpredictable problems and weaknesses, a set of trained and tested backup individuals or other detection devices or equipment should be immediately available to serve as a substitute in the event of a dog's illness or other sign of abnormal behavior. A dog should be used only if it can be shown to detect firearms or explosives with equal or greater confidence than existing alternatives. A dog may be particularly useful in the search of vehicles or oversize packages. As the duration of a dog's effectiveness for performing search functions may be limited, it should be used only as a secondary aid. Because of the "intimidation factor," the use of a search dog can provide an advantage during searches conducted with potentially uncooperative subjects.

Canine detection is also commonly used in real-world explosive detection applications. Dogs surpass current manmade technologies in terms of their high degree of mobility and their ability to follow a scent rapidly to its source. Thus, canines excel at applications with a significant search component. Dogs can, in principle, be trained to detect any explosive, and typically a dog might be trained on 10 to 20 different explosive scents.

The main drawback of canine detection is the dog's limited duty cycle. Depending on its training regimen, a dog can work from one to several hours without a break, but it cannot work on a 24/7 basis like a piece of manmade equipment that is functioning as intended. Those desiring canine detection can either purchase trained dogs or obtain the service from a qualified vendor. It is worth noting that dogs must constantly maintain a training schedule, usually tens of hours per month, to maintain their detection capabilities at a high level. The dog must also normally work with a single, dedicated person called the handler. The handler and the dog form a team that provides the detection capability. Canines are excellent for vehicle screening and for searches of buildings or other facilities. For a variety of reasons (e.g., some people fear dogs, the liability issue if someone is bitten), dogs are not normally used to screen people.

2.3.6.4 Neutron-Based Bulk Detection Techniques

Among bulk detection techniques (other than X-Ray imaging) perhaps the most common are probing neutron-based techniques, such as thermal neutron activation (TNA) and pulsed fast neutron analysis (PFNA). Thermal neutrons have energies of approximately 0.01 electron volt, while fast neutrons have energies greater than 5 to 7 million electron volts (MeV). In these techniques, the object to be interrogated is irradiated with neutrons, which induce certain atomic nuclei in the irradiated materials to emit gamma rays of characteristic energies, which are then detected. In explosive detection applications of TNA, one looks for gamma rays emitted by nitrogen nuclei to identify nitrogen-rich materials, which may be explosives. In PFNA, gamma rays from other low-mass nuclei, including carbon and oxygen, can be detected. The PFNA technique thus provides more information than TNA and can also identify materials with more certainty, but only at a much greater cost.

These techniques are normally utilized to search for large bombs in vehicle cargo holds or cargo containers, and a typical limit of detection might be 50 to 100 pounds (22.7 to 45.4kg). Screening people with these techniques is prohibited because of health concerns. Since information is provided on atomic composition, these techniques have varying degrees of material specificity, which are lacking in a simple X-Ray image. The main drawback of these techniques is their expense. The techniques are also relatively slow; screening a large vehicle may require several minutes.

3. TESTING OF SEARCH EQUIPMENT AND METHODOLOGIES

3.1 <u>Introduction</u>

Proper testing of equipment and detection methodology is essential to ensure that the screening performed at access control portals achieves its intended purposes. An overview of the topic requires the consideration of the different types and aspects of testing and of specific types of detection technologies.

General types and aspects of testing that are important include the following:

- instrument calibration

- operability testing

- performance testing

- statistical aspects of testing

- test documentation

- maintenance of equipment

After reviewing these topics, this section discusses testing for the different types of detection equipment and methods discussed in this report, which include the following:

- access control equipment, including biometrics and card reader/PIN devices

- metal detectors

- X-Ray-based imaging equipment

- CT scan systems

- trace explosive detection equipment

- canine explosive detection

Since all of these types of equipment and detection methods differ, there are differences in how to conduct testing. However, several key questions are important to ask when considering the testing of any type of detector:

- How should the equipment be properly calibrated?

- What instrument settings should be used during the testing?

- What are appropriate test objects?

- How often should the equipment be tested?

- Is the test properly documented for future reference through written test procedures and written information that records the relevant test conditions and results?

3.1.1 Instrument Calibration

Proper instrument calibration is vital to ensure that any type of detection equipment is functioning as intended. Calibration procedures vary widely based on equipment type. Some equipment will need to be calibrated in the field at least on a daily basis, while other equipment may require only infrequent calibration to be performed by the manufacturer. Commercial equipment purchased from reputable manufacturers should have a detailed user's manual that provides information on calibration and appropriate calibration procedures, if needed. Manufacturer or vendor training in the use of the equipment may be useful and could be a part of the purchase contract, and this should include training in equipment calibration (where applicable). In such cases, manufacturer or vendor training should continue until the site personnel responsible for operation and calibration of the equipment have performed calibration and feel comfortable with the procedures. In addition to any regular calibration that may be recommended by the manufacturer, equipment generally needs to be calibrated when it is moved to a new location, put into use again after a long period of inactivity, has any instrumental settings changed, or appears not to be functioning properly.

Instrument calibration establishes that the instrument is functioning accurately within certain parameters. Normally, the instrument is tested with a test object that it should be able to detect and the response is measured. The response may be quantitative or qualitative and single or multiple tests can be involved. Calibration is not always performed with an actual threat item. For example, trace explosives detectors are often calibrated with microscopic quantities of explosive materials that are much too small to be a threat, but these tiny masses are something that the instrument should be able to detect if it is functioning as intended. Calibration is the most basic type of test in that a piece of equipment needs to be calibrated correctly before operability or performance testing can take place.

3.1.2 Operability Testing

Operability testing needs to be conducted frequently and involves testing access control equipment and search equipment to confirm that the equipment is functioning and performing within designed capabilities. Operability testing usually differs from instrument calibration in that the item or items used to challenge the system are normally threat items or mock threat items not provided by the manufacturer as part of the standard instrumentation suite. Most physical protection programs rely highly on ensuring that unauthorized items are not introduced into protected areas; therefore, the operability testing of search equipment should be conducted at the beginning of each security duty shift and tested for performance at least once during each 7-day period. Since operability testing normally involves rather quick and straightforward tests, performing it frequently does not constitute an undue burden to site operations. Note that a detector is likely to fail operability testing if it is not calibrated properly. The following sections provide examples of possible operability tests for different types of detection equipment.

3.1.2.1 Biometric Device

A person can attempt to pass through the checkpoint using someone else's identity. For example, person X could claim to be person Y. If the system is working correctly, the attempt to gain access should be rejected because the biometric feature presented by person X will not match person Y's template.

3.1.2.2 Portal Metal Detector

A person carrying a metal knife or gun can pass through the portal and the metal threat object should be detected.

3.1.2.3 X-Ray Imaging Equipment

A person can conceal a weapon or explosive simulant in a briefcase or other hand-carried item. The test can usually be made more or less difficult by adjusting the amount of clutter (i.e., other items besides the test object) inside a briefcase. A test can also be performed using threat image projection, where the image of a threat item is projected onto the image of a screened item.

3.1.2.4 Trace Explosives Detection Portal

A person who has been deliberately contaminated with trace explosive material can walk through the portal.

3.1.2.5 Canine Screening at a Vehicle Access Control Portal

If a dog is used to screen entering vehicles, place an appropriate amount of explosive material in a vehicle to determine whether the dog detects it.

Clearly, many tests can be devised. Keep in mind that the basic goal of operability testing is to confirm that the equipment properly detects one or more items that it should be capable of detecting once it is turned on and calibrated. For this reason, operability testing is normally conducted with the full knowledge of the equipment operator.

Operability testing may be limited by safety concerns. For example, at many locations it is not possible to carry out tests with threat quantities of real explosives. In such cases, trace explosive detection equipment can be challenged using trace explosive contamination, and bulk explosive detection equipment can be challenged using an appropriate explosive simulant. Safety concerns may also dictate that weapons used to challenge metal detectors or X-Ray imaging equipment are unloaded or disabled.

3.1.3 Performance Testing

Performance testing is more detailed and systematic than operability testing. Though it is conducted less frequently than operability testing, performance testing should be conducted at least once every 7 days. The purpose of performance testing is to document as completely as possible how well a piece of detection equipment functions over a specified time period under real-world operational conditions within its intended range of function.

Test objects used in performance testing can be the same or similar to those used in operability testing. However, performance testing will generally involve a broader suite of test objects. Utilizing a diverse set of threat objects in this type of testing is desirable because, if only one or a few test objects are used, operators may narrowly focus on only those test objects while ignoring other, equally valid potential threats. A diverse set of test objects can also validate that the equipment can detect across its intended range of function. It is important that performance testing on detection systems consists of testing the system across its entire intended range of function as specified by the system manufacturer. Most detection equipment manufacturers offer test kits or specific test objects that challenge the detection capabilities of the detection system through the entire range of threat objects the system is designed to detect. Many test objects, such as explosive trace test objects, have a shelf life. Therefore, it is important to ensure that these test objects have not exceeded the specified shelf life since use of outdated test objects will not provide accurate results.

Performance data produced in performance tests should be treated as sensitive unclassified information, and access to such data should be denied to personnel who do not have the appropriate access authorization and a need to know.

3.1.4 Statistical Aspects of Testing

One goal of performance testing is to calculate a probability of detection (P_D). The P_D is a number varying from zero to one or from 0 to 100 percent that applies to a specific detector under a specific set of test conditions. Calculation of P_D can be performed after a sufficient number of tests have been performed in order to gain statistical confidence. Note that the test conditions must be clearly defined if the P_D value is to have any meaning. For example, the P_D for a trace explosive detection system will depend on the type of explosive material being detected; how that material is presented to the piece of equipment (e.g., contamination on a person's shirt, bulk explosive hidden in a briefcase); the instrumental settings being used; and possibly environmental conditions, such as temperature, pressure, and humidity. Changing any of these factors or conditions may affect the P_D value. Some factors will have a major impact while others will have only a minor impact.

The discussion that follows in this section is based on statistical theory. It is equally relevant to all detectors and detection methods. When a P_D is reported, it is normally reported along with a confidence level (e.g., a 90 percent probability of detection with 95 percent confidence). This means that it can be stated with 95 percent confidence that the probability of detection is at least 90 percent. The probability of detection that needs to be demonstrated for a piece of detection equipment depends on the type of equipment and the desired confidence level. The following paragraph focuses on 90 percent probability of detection with 95 percent confidence as an illustrative example.

Table 1 lists the total number of tests, the number of successful tests, and the corresponding maximum number of unsuccessful tests that allow a P_D of 90 percent to be claimed with 95 percent confidence. This claim can be made if 30 out of 30 tests successfully detect a test object. If even one test is unsuccessful, the claim cannot be made based on 30 tests. Rather, the number of tests must be increased to 40, and if 39 out of 40 tests are successful, a P_D of 90 percent can again be claimed with 95 percent confidence. Similarly, the same claim can be made if 48 out of 50 tests are successful. If, at any point in a test series consisting of a maximum of 50 tests, the total number of unsuccessful tests reaches three, then the test series has not succeeded in establishing a P_D of 90 percent with 95 percent confidence. In that case, the test series should be halted and the system operation reviewed for any obvious problems, such as incorrect instrumental settings. If the operators cannot identify any such problem, it is recommended that an onsite alarm technician, the manufacturer, or the vendor be consulted, as appropriate. If the manufacturer or vendor cannot identify a problem, it may be concluded that the system is operating optimally and that it is not capable of achieving a 90 percent P_D with 95 percent confidence for the test in question. The system might still be retested later to verify that the test result is reproducible.

Table 1: Total number of tests and number of successful tests required to establish a P_D of 90% with 95% confidence

Total Number of Tests	Number of Successful Tests	Number of Failures
30	30	0
40	39	1
50	48	2

For an arbitrary number of attempts, N, the required number of successful detections, n, to establish a detection probability, p, with confidence, C, is dictated by the cumulative binomial distribution.

$$C = \sum_{n=0}^{N} \binom{N}{n} p^n (1-p)^{N-n} \tag{1}$$

where $\binom{N}{n}$ is the binomial coefficient

$$\binom{N}{n} = \frac{N!}{(N-n)!\,n!} \; . \tag{2}$$

The equations can be used to determine the minimum number of successful attempts, n, required to establish a given detection probability, p, with confidence, C. Alternatively, the binomial distribution can be used to determine the actual detection probability, p, for a given number of attempts, N, and confidence, C. Consequently, these equations can be applied to determine the system's actual detection probability at several confidence levels from accumulated test results.

3.1.5 Test Documentation

Whether the testing performed is simple calibration, operability, or performance, it needs to be documented. The importance of this documentation cannot be overstated. The data will likely be useless if the test results are subsequently reviewed and it cannot be determined what was tested, under what conditions the test occurred, and the nature of the results. While written calibration procedures are normally provided in the equipment user's manual, the user should develop or obtain separately detailed written procedures for operability and performance testing, and these procedures should be rigorously followed. All test data should be recorded in ink on a template form developed for the particular piece of equipment used.

The facts that are recorded should include, but are not limited to, the following:

- the system being tested (i.e., name, make, type, model number, serial number)

- all instrument settings used during the test

- the system operator

- any other site personnel involved in the test, such as an alarm technician

- date and time of the test

- the threat object, mock threat object, or other item being used to challenge the detector

- how the threat object or other test object was concealed or transported (e.g., hidden under clothing, in a cluttered briefcase)

- whether or not a detection is made

- environmental conditions, such as atmospheric pressure, temperature, and humidity

- If possible, evaluate the probable cause when no detection occurs. Such causes may include system malfunction, improper or no calibration, or use of a test object that the equipment is simply not capable of detecting.

File completed test documentation in a secure location. All data can be entered into a computer database and backed up on a regular basis. This protocol will allow data to be recovered if the original form is destroyed or lost.

3.1.6 Maintenance of Equipment

Proper maintenance of equipment is important to ensure that the equipment is functioning as intended. However, maintenance requirements vary widely depending on equipment type; therefore, the following sections discuss this topic for each type of detection equipment. For any piece of commercial equipment, it is important to obtain information on maintenance. Maintenance instructions, which are usually provided by the equipment manufacturer or vendor, should be followed as precisely as possible. The complexity of maintenance procedures is highly variable. Some procedures can be performed routinely on site by equipment operators, while other procedures may require a site visit by the manufacturer or vendor; other procedures may require shipping the equipment back to the manufacturer or vendor.

3.2 Personnel Access Control Devices

3.2.1 Test Objects for Access Control Devices

Tests for personnel access control devices require a mix of people, procedures, and hardware. The test objects typically involve a badge, PIN, and biometric template for some claimed identity. Each such identity that is tested will be used by at least one individual to test a feature of the access control system. Tests should be designed to test all features of the access control system.

These test objects may allow the holder to gain access to sensitive areas and should be handled with appropriate security precautions. Results of these tests should be protected to avoid inadvertently disclosing vulnerabilities of the access control system. In particular, care should be taken to ensure that the actual identity of a person whose biometric template is used in testing is not revealed. Revealing this information may allow volunteers to learn that they may be able to pass for other individuals authorized to enter the facility. Also, keep in mind that an access control PIN is often sensitive unclassified information and thus must be protected from disclosure to unauthorized users.

3.2.2 Configuration

Modern access control systems offer a great deal of flexibility in configuration. With this flexibility comes the opportunity for setup or configuration errors. This flexibility also makes it impossible to define a universal testing protocol that can be applied in all situations. Examples of common configurations are included here but should not be considered a complete set.

3.2.3 Calibration of Personnel Access Control Devices

Typically, the only personnel access control device that requires calibration is a biometric reader. The first step in the calibration process is to ensure that the reader is acquiring an accurate representation of the biometric feature being measured. This step should be performed according to the manufacturer's specifications and recommended frequency.

The second step that may be required for calibration is adjusting the acceptance threshold. Usually, this should be set as recommended by the manufacturer. Once the biometric system is configured, it should be tested as described below.

3.2.4 Operability Testing of Personnel Access Control Devices

Operability testing of personnel access control devices should involve challenge tests to confirm that these devices are functioning as intended. The tests should confirm that properly enrolled personnel are allowed access and that unauthorized personnel are not allowed access. The first type of test is to observe the system functioning in normal operations. Observing properly enrolled personnel entering the facility during the workday verifies the operability of the system and can serve as an operability test. If the false rejection rate is substantially higher than it should be, this will quickly become evident based on the number of properly enrolled people who are having trouble entering the facility or area. The main concern of operability testing then becomes ensuring that access is being denied when appropriate. An easy way to test this is to deliberately simulate an unauthorized entry. This can be done by using the normal entry procedure and entering the wrong PIN or using one person's credential and another's biometric identification. This test should deny access in all cases.

3.2.5 Performance Testing of Personnel Access Control Devices

Performance testing should be designed to challenge every feature of the personnel access control system to ensure the features are working as intended. The tester will need to develop a test plan based on the security needs of the specific facility. For each of the questions listed under the "Areas to be Tested" column, the tester will need to set up the appropriate test. For example, a facility that has an access control system with two security areas that uses badges and PINs would need to either create a badge for each of the areas or use existing badges to test the area feature. Each badge should allow access to the appropriate area and not allow access to the other area. An exception to the "not allow" test would be if one area is a subset of the other. In this case, the badge for the area that makes up the subset should allow entry to both areas. Each badge should be tested with the correct PIN and with a different PIN. Each PIN should also be tested to ensure that it allows access only during the correct hours and dates and that it does not work after its access rights expire.

Table 2: Testing areas for personnel access control systems

Feature	Areas To Be Tested
Access Control by Area	Does the system allow a tester into the appropriate area? Is a tester prevented from entering inappropriate areas? If multiple areas exist at each security level, it is sufficient to test one area at each level.
Schedule (by the clock and by the calendar)	Is the tester allowed in during scheduled hours? Is the tester prevented from entering during nonscheduled hours? Is the tester prevented from entry before the appropriate calendar day? Are all entries prevented after access rights expire?
Personal Identification Numbers	Will entering the correct PIN allow entry? Will entering the incorrect PIN deny entry?
Logging Attempts, Both Successes and Failures (documenting access)	Are all access attempts logged appropriately?
Alarm on Multiple Failures*	Will the system generate an alarm that is received at the alarm stations upon multiple failed attempts at entry?
Lockout on Multiple Failures*	Will the system lock a badge out of the system or require a time-out period to expire after multiple failed attempts at entry?
Antipassback*	Will the system prevent a tester from reentering a security area without logging out?
*If the feature is present in the security system	

Any failures during this testing should be analyzed to determine the cause. Any failure that cannot be attributed to operator error or an error intrinsic to the biometric system should be analyzed to determine its root cause and the system should be repaired to eliminate the cause of the error. Errors that could be attributed to humans include incorrect swiping of a magnetic strip badge, incorrect entry of a PIN, and incorrect presentation of the biometric characteristic or not operating the biometric reader correctly. Errors intrinsic to the biometric system include false acceptance and false rejection errors. The false acceptance and false rejection errors should be analyzed to determine if they fall below the acceptable rate designated by the facility.

Performance testing of the personnel access control system should occur when the system is installed or a system modification occurs. System modifications include new software revisions introduced to the system, addition of controlled access points or areas, or the repair or replacement of hardware. In addition, the system should be performance tested periodically to ensure that it continues to operate as intended.

3.2.5.1 False Rejection and False Acceptance Rates

In performance testing of personnel access control devices, it is particularly important to determine the false reject and false accept rates. For false rejection testing, a group of individuals will need to be enrolled in the biometric system. These individuals will attempt to have their identities verified at an access control portal using the normal procedure. The data from this test should be analyzed as discussed in Section 3.1.4, "Statistical Aspects of Testing." In this case, each verification attempt is considered a test, and each successful verification is considered a successful test. The goal of each test is to determine the probability of granting access appropriately.

For false acceptance testing, individuals who have successfully used the biometric system will attempt to claim a different identity that has been enrolled in the biometric system. This is one type of test that may employ the use of test badges. For example, a badge may be prepared that is encoded to call up the biometric template of "John." This badge may be issued to "Joe" for testing purposes to determine if "Joe" can pass for "John." In this test, each verification attempt is considered a test, and each successful rejection is considered a successful test. The goal of this test is to determine the probability of detecting an imposter. Refer to Section 2.4.4 for guidance on data analysis.

A minimum accuracy of 90 percent at a 95-percent confidence level should be achieved in false rejection and false acceptance testing. In both types of testing, it is important to use a diverse set of individuals as test subjects. No one individual should participate in more than 10 percent of the tests. The claimed identities for false acceptance testing should be randomly selected from the general user population. At a minimum, this type of testing should be conducted annually. The testing could be performed in a single day or over a period as long as a month. Longer test performance periods are not recommended because gradual changes in system performance may render the statistical analysis of the results invalid.

3.3 Metal Detection

3.3.1 Principles of Operation

Metal detectors are employed to detect weapons and explosive or incendiary devices containing metal parts upon entrance to protected areas and material access areas and for detection of shielding materials upon exit from material access areas. There are historically two general types of walkthrough (portal) active metal detectors with only one still in use today. The earliest walkthrough (portal) active metal detectors (still in wide use for hand-held instruments) were of the continuous-wave type. After the introduction of pulse metal detectors for walkthrough portals, this new technology rapidly replaced continuous-wave technology for portal applications. Regardless of the specific technology, all active metal detectors rely primarily on the fact that a time-varying magnetic field induces currents in metal which can be detected. Metal detectors are widely used in security applications because of their high throughput rate, effectiveness at detecting weapons, and nonintrusive nature. Refer to Section 2.3 for additional information on the operational principles of metal detectors.

3.3.2 Factors Affecting Metal Detector Operation

3.3.2.1 Static Metal

Most metal detectors have the ability to ignore static metal. This feature permits them to be positioned near stationary metal objects without alarming constantly. However, nearby metal structures can still affect the sensitivity of a metal detector and, whenever possible, a metal detector should be located at least 3 to 4 feet away from metal structures. Ferromagnetic structures, or materials with a relative permeability[1] much greater than one, can distort the transmitted magnetic field, disrupting the uniformity of the detection field. Nonferromagnetic metal structures can act as shields, reducing field strengths. Metal in the floor should also be considered when installing a metal detector. An elevated ramp or aluminum shield may be necessary to offset the effect of metal structures in the floor. Static metal can also provide a path to electromagnetic noise. Care should be taken to use proper grounding and isolation techniques to prevent a metal structure from conducting noise from a far-away source to the metal detector.

3.3.2.2 Installing Near Other Equipment

Installing multiple metal detectors in close proximity to, or operating a metal detector in conjunction with, X-Ray equipment may require special installation procedures. Consideration for ensuring the detector is provided adequate space will likely reduce the amount of interference caused by other search equipment located in the vicinity of the metal detection device. Consultation with the equipment manufacturer before installation can provide techniques to minimize the effects of interference from these sources.

3.3.2.3 Hand-Held Metal Detector Interaction with the Floor

Similar to the case of walkthrough detectors, metal in the floor can be a problem with hand-held metal detectors. The hand-held detectors, with their high sensitivity, can alarm when passed close to a floor that contains metal (e.g., reinforcing bars). If the interference is minimal, a simple solution is to make sure that the paddle or loop of the detector is held vertically because the edge of the detection coil is the least sensitive part. Holding the coil vertically will minimize the response to metal in the floor. If interference is still a problem, a wooden platform may be constructed on which the person being searched can stand. The construction of the platform should make use of glue and other nonmetallic fasteners to prevent false alarms caused by metal nails or screws.

---

[1] Magnetic permeability is a property of all space and is influenced by the material present in that space. Permeability is a proportionality constant that relates the strength of a magnetic field within a volume of space or material to the magnetic field surrounding that volume of space or material. The International System of measurement (SI) unit of magnetic permeability is based on inductance per unit of distance and is given in henries (the SI unit of inductance) per meter. The permeability of free space (vacuum or air since air does not significantly change the permeability) is approximately 1.257×10^{-6} henries per meter. Relative permeability is the magnetic permeability of a material divided by the magnetic permeability of free space. The impact of magnetic permeability is that materials with high relative permeability will greatly strengthen the magnetic field within the material as compared to the magnetic field in the air surrounding the material.

3.3.2.4 Site Selection

The site of a high-sensitivity metal detector should be selected carefully, and the power source for the metal detector should be relatively free from noise or power surges. The presence of any metal objects in the immediate vicinity should be considered for every metal detector installation, but especially when greater sensitivity is required. The presence of steel or other metal in the floor, walls, or ceiling of a potential site should not be overlooked. A high-sensitivity metal detector should be located away from all metal objects.

3.3.2.5 Properties of an Object That Affect Metal Detector Response

Metal detectors use eddy currents and their associated magnetic fields to detect the presence of a conductive material. Two properties, the geometry of the target and its electrical and ferromagnetic properties, can affect the magnitude of the eddy currents and therefore the magnitude of the metal detector's response.

3.3.2.5.1 Target Size

The geometry of the target influences the magnitude of the eddy currents in several ways. First, the cross-sectional area normal to the magnetic field is proportional to the induced voltage within the target; therefore, objects that have the same cross-sectional area have nearly the same induced voltage. If metal detectors simply responded to the induced voltage, objects with the same induced voltage would produce the same response. However, metal detectors respond to the eddy currents that result from the induced voltages. In accordance with Ohm's Law, the resulting current is proportional to the voltage and inversely proportional to the resistance.

3.3.2.5.2 Target Shape

The shape of the target is a major factor in determining the magnitude of the resistance of the conductive path of the target and therefore the magnitude of the eddy current. A given material has a fixed resistivity so the resistance of a target's conductive path is the resistivity of the material multiplied by the path length. The length of the electrical path is effectively the length of the perimeter of a cross-sectional area of the target that intersects a plane normal to the magnetic field. A useful example is to compare a circle and a rectangle, both with an area of 9 square inches. If the rectangle has a length of 9 inches and a width of 1 inch, the rectangle's perimeter is about twice as long as the circle's. Even though the induced voltage of each target is the same (because of equal area and Faraday's Law), the resistance of the rectangle is more than twice that of the circle. The result is that the circle will produce an eddy current that is over twice that of the rectangle and will thus be easier to detect. Targets of complex shape will tend to produce smaller responses in a metal detector. Complex, asymmetrical shapes also cause variation in response because of orientation by presenting different cross-sectional areas to the magnetic field.

3.3.2.5.3 Target Composition

The types of metals and other materials that comprise an object can strongly impact the ability of a metal detector to detect that object. Nonconductive materials do not contribute to detection. For some firearms, the grips and a significant portion of the frame are composed of nonconductive materials like plastics. In the case of these firearms, only the metallic parts contribute to the signals that allow the detector to detect the firearm. Various metals have different degrees of conductivity (the inverse of resistivity). Modern weapon manufacturers use a wide variety of metals, including aluminum, zinc alloys, various alloys of stainless steel, and the more common carbon steel. These materials have different degrees of resistivity and some, such as aluminum and zinc, are not ferromagnetic. As discussed earlier, these factors impact the magnitude of the induced eddy current and thus impact the detectability of the firearm.

3.3.2.5.4 <u>Uniformity of the Detection Field</u>

A metal detector is the combination of transmitting and receiving systems (Figure 8). The transmitting and receiving antennas are usually on opposite sides of the portal walkway. The transmitter generates an electromagnetic field in a pattern determined by the configuration of the transmitting coil or antenna. This field varies in strength as the inverse square of the distance from the antenna. Since the field strength falls off as the distance from the transmitter increases, the magnitude of the induced eddy current also falls off as the distance from the transmitter increases, suggesting that the area of lowest sensitivity would be on the receiver side of the detector. However, as the target is passed through towards the receiver side of the detector, its close proximity to the receiver allows its smaller signal to be detected more easily. The result of these two competing effects, which has been confirmed experimentally, is that the area of lowest sensitivity is near the middle of the detector archway. Another field factor that affects uniformity is the fringing of the magnetic lines of flux at the extremes of the transmitter coil. The fringing of the lines of flux causes areas of lower sensitivity at the top and bottom of the detector.

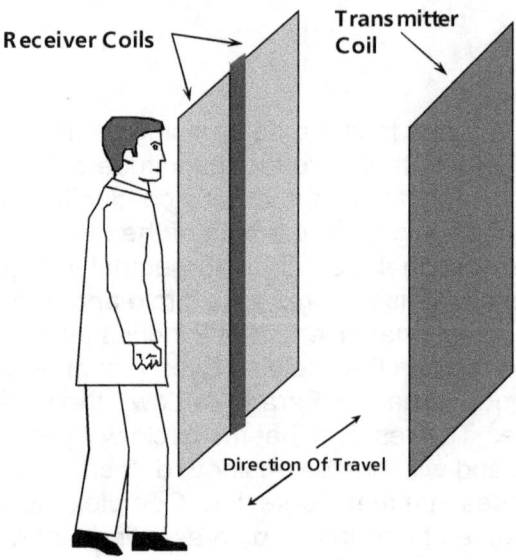

Figure 8: Schematic representation of metal detection portal

3.3.3 Test Objects for Metal Detectors

It is important that the capabilities and vulnerabilities of metal detection equipment are identified and accounted for before system implementation. Identifying and accounting for the potential vulnerabilities of detection equipment used for the conduct of searches provides assurance that the access control and search processes support the specific design goals of the overall physical protection program of the facility. It is recommended that metal detection portals for personnel and package exit screening are tested using test objects that are as realistic as possible. Test objects should be consistent with the type of objects that pose a threat to the area being protected.

Metal detector test objects should include (at a minimum) handguns, and care must be taken in selecting the firearms or test objects to be included in the test kit. The firearms or test objects chosen should account for the worst case of firearms that can represent threats to the areas the metal detector will protect. A variety of firearms or test objects should be selected based on variables such as size, shape, and composition. One possible test kit would include a small firearm composed mainly of aluminum (high conductivity, nonferromagnetic), a small firearm made of carbon steel (relatively high conductivity, ferromagnetic), and a small stainless steel firearm (low conductivity, ferromagnetic). This set contains the minimal number of firearms that represent a fairly wide spectrum of modern weapons.

Using firearms for testing metal detectors has two drawbacks. First, it is important to make sure the firearms cannot be used as an actual threat. Typically, the firearms used for metal detector testing are rendered unusable as weapons. Firing pins are filed short or removed, and the firearms may be filled with plastic materials such as epoxy to prevent them from being loaded. Another possible solution is that, after the firing pin has been made inoperable, the firearm can be encapsulated in a plastic material like urethane that is brightly colored and may be in the shape of a block to prevent the test object from even looking like a firearm. The second problem is that in today's firearm market, the selection of specific models is constantly changing. A specific firearm model selected as a test object may not be commercially available 2 or 3 years later. This can make the generation of a new test set difficult. Some organizations have implemented bulk purchase of test firearms which are then stored as required for future use.

3.3.3.1 Alternative Test Objects for Metal Detectors

Various organizations have employed alternative test objects to avoid the drawbacks of using firearms. These objects are fabricated to simulate a small weapon. One notable example is the operational test piece (OTP) that has been used at airports to test metal detectors. The OTP, which was developed by the Federal Aviation Administration (FAA), is constructed to be similar in composition, size, and shape to some commonly available compact firearms. While the OTP is similar in size and is composed of a common alloy, it does not always cause the same response in metal detectors. The complexity of the firearm (the fact that it is composed of many smaller parts) caused differences in the signal signature. The Marshal's Service employed a similar test piece to test courthouse metal detectors, but the results were similar. Currently, many alternative test objects for metal detection equipment that meet the specifications of the FAA for testing metal detectors are in use. It is important during selection of test objects that the facility conducting the test be aware of and account for the differences in response that its particular metal detectors have on the test objects that it intends to use.

There generally will be no alternative test object for material access area exit applications since the test object that represents the real shielding object is fabricated for this purpose and would not differ from the alternative.

3.3.4 Calibration of Metal Detectors

Metal detectors must be calibrated periodically. Metal detectors are very stable in operation unless something changes, such as the operational environment (i.e., elevation, humidity, and temperature), additional metal objects in proximity to the unit, or the detector settings. Therefore, metal detectors must be calibrated when the environment has changed and an operability or performance test detects changes in performance or when any of the settings have been changed. Calibration should also be performed if the detector starts having problems or following any maintenance. Finally, the manufacturer may recommend calibration at periodic intervals even if no changes occur. If this is the case, calibration should be performed at least as frequently as recommended by the manufacturer.

3.3.4.1 Calibration Adjustments

In general, calibration of metal detectors is achieved using the test kit appropriate for the application for which the metal detector will be used and following all instructions provided by the manufacturer. The detector should be installed at the location where it will operate. The test kit is used to find the minimum sensitivity setting required to reliably detect all test objects in all orientations anywhere in the detector's detection field. Once this setting is determined, an operability test can be conducted to see whether the system passes. If it passes, it can be put into operation. If it fails, then the sensitivity is adjusted higher and the operability test is repeated. This process continues until the detector passes the operability test and is placed in operation.

3.3.5 Operability Testing of Metal Detectors

Metal detectors should be operability tested at the beginning of each security duty shift. The operability test will normally involve taking a test object through the portal once to verify that an alarm is recorded. The test object used can be an object in the instrument test kit or even the sidearm or flashlight of a security officer who is present.

3.3.6 Performance Testing of Metal Detectors

Performance testing should be conducted regularly to confirm that the system will detect metal with acceptable sensitivity. In general, the system should be capable of detecting metal with a probability of 85 percent (see Table 3) with 95 percent confidence. Performance testing to establish this probability of detection should be conducted at least once every 7 days. Tests should also be conducted following repairs or modifications to the system, following changes in its operating environment (including the addition of nearby metal objects), or following system calibration.

A minimum of 19 attempts should be conducted for performance tests and tests following repair, modification, or calibration. As Table 3 indicates, for 20 attempts, 20 detections are required to confirm with 95 percent confidence that the system has an 85-percent detection probability. Actually, 19 of 19 are all that are required with a no-miss testing strategy. If the first 19 attempts do not yield 19 successful detections, the test can be extended to 40 attempts. If the system detects the target in fewer than 38 of the cumulative 40 attempts, the system's operational status should be checked for an out-of-calibration condition. If the system is out of calibration, it should be recalibrated according to the manufacturer's instructions, and the operability test should be repeated from the beginning. If the system still fails to pass 38 of 40 attempts following recalibration, the system should be removed from operation, pending manufacturer repairs.

This set of attempts should be conducted with the most difficult to detect test object that is part of the instrumental test kit, with the test object at the most difficult location for detection within the portal, and with the test object in the most difficult to detect orientation. The persons taking the test object through the portal should proceed at a normal walking speed of approximately 1 meter per second.

Table 3: Minimum number of detections required to establish an 85% detection probability with different confidence levels

	Minimum Number of Detections		
Attempts	50% Confidence	70% Confidence	95% Confidence
5	5	NA	NA
10	10	10	NA
15	14	14	NA
20	18	19	20
40	35	36	38
60	52	53	55
90	77	79	82

3.3.6.1 Other Variables in Performance Testing

Performance testing of metal detectors may also investigate other variables. It is recommended that the manufacturer be consulted in designing a detailed set of performance tests of this type. Test parameters that can be investigated for a portal metal detector include the following:

3.3.6.1.1 Variation of the Test Object

At a minimum, the set of test objects should include three types of small firearms as described previously, and the portal should be tested with each test object.

3.3.6.1.2 Variation of the Test Object Location within the Portal

As described previously, the portal sensitivity varies with both the height and lateral placement of a metal object within the portal. Performance testing should include tests performed at a detailed grid of points to map out the portal sensitivity. This can be done manually by simply holding the test object at different locations within the portal as someone walks through or by having the test object attached to different locations on the person's body. A more sophisticated approach is to have a nonmetallic rack with numerous compartments (similar to a rack for wine bottles) that fits inside the portal. A test object can then be attached to a nonmetallic handle that is several feet long and passed through the various compartments of the test rack. This approach ensures that the test object is passed through the metal detection portal at a well-defined set of reproducible locations. This is sometimes referred to as a "honeycomb test" because the numerous open compartments within the rack resemble a honeycomb. Research is currently being done to develop robots that can pass test objects through a portal at precisely defined location, but these robots are not yet widely available.

3.3.6.1.3 Variation of Test Object Orientation at a Given Location within the Portal

As discussed previously, this can make the test object easier or more difficult to detect.

3.3.6.1.4 Variation of Walking Speed

Portal metal detectors are designed to detect metal best at a certain optimal walking speed defined by the manufacturer. Performance testing should investigate faster and slower walking speeds to determine how severely the sensitivity falls off as this parameter is varied.

3.3.6.1.5 Variation of the Operational Environment

Every effort should be made to make the normal operating environment as optimal as possible by limiting interference from both static and moving metal in the general area. Performance testing might include changes in this environment, such as moving another piece of equipment closer to the metal detector to see how much this negatively impacts detection capabilities.

3.3.6.1.6 Nuisance Alarm Testing

To understand the system's nuisance alarm rate, it is useful to note how often the portal alarms on small metal objects that pass through it, such as belt buckles, watches, and coins. Part of the data on such alarms can be generated by noting how often the portal alarms during screening operations. However, noting the numbers of alarms during screening does not provide any information on how often such small objects went through the portal without producing an alarm. Obviously, the nuisance alarm rate will vary as the portal sensitivity is adjusted.

3.4 X-Ray Inspection Systems

3.4.1 Operational Principles

X-Ray imaging systems provide a means to inspect the contents of packages without opening the packages and sorting through their contents. When used by well-trained operators, these systems can provide a way to rapidly examine packages that can be more thorough than a manual search. One of the main advantages, other than speed, is that these systems can image the interiors of innocuous-looking objects and reveal hidden compartments that may otherwise be overlooked.

3.4.1.1 Package Search X-Ray Imagers

The interaction of X-Rays with materials falls into one of two categories: transmission and backscatter. Both are used in package screening. Transmission is concerned with the X-Rays that are not absorbed and are transmitted through the package. Backscatter refers to X-Rays reflected from a material. In the case of backscatter, the X-Ray source and X-Ray detectors are on the same side of the package. The degree or proportion that a specific energy of X-Ray is transmitted (or, conversely, absorbed) can reveal something about the nature of the material being inspected. Likewise, the proportion of the X-Ray energy that is backscattered can also reveal something about the nature of the material. In both cases, the X-Ray energy signal can be used to construct an image of the structure and contents of packages.

3.4.1.2 Package Search System Description

X-Ray baggage inspection systems like those used in airports constitute the major commercial package search systems available and in use today. These systems, or adaptations, are also employed in other facilities, such as nuclear power plants, correctional institutions, Government facilities, and industrial facilities. Most commercial X-Ray baggage inspection systems are designed for high throughput of handbags, briefcases, and baggage carried by airline passengers. An image of the contents of a package is obtained by various X-Ray techniques. The image depends on the density and transmission or backscatter characteristics of articles within the package. The resulting image is displayed directly or processed on a computer and then displayed on a monitor. The image is viewed by a human operator and detection is accomplished by visual pattern recognition. Training of operators in visual pattern recognition is extremely important for successful system operation.

A typical package search system consists of the following:

- A radiation source

- An X-Ray sensing system

- An image processing system

- A display

3.4.1.3 X-Ray Sources

Sources of penetrating radiation for package search systems include conventional X-Ray tubes. Standard X-Ray tubes that operate at approximately 65 to 200 kilovolt potential (kVp) are suitable for package search applications, with the vast majority of small package systems using voltages near 65 kVp.

To protect camera film and other radiation-sensitive materials, some means of reducing the total exposure (dose) is needed. Various dose reduction methods have been developed, such as pulsed sources, fan beams, and flying spot. The fan beam exposes the package to a narrow fan of radiation as the package is carried past the source on the conveyor. The flying spot method carries the process a step further by exposing the package to a small-diameter pencil beam that is scanned raster fashion across the package as it is carried past the source. The pencil beam is formed when a small portion of a collimated fan X-Ray beam passes through the radial slits of a spinning chopper wheel, producing a small area spot beam, which sweeps along the package. Also important in dose reduction is the use of highly efficient sensors and image processing, which has allowed the use of lower X-Ray intensities, resulting in ever smaller energy doses.

High-energy radiation (i.e., greater than 1 MeV) is needed to penetrate dense material that could conceal contraband in large containers and vehicles. Systems utilizing these energies are employed to examine large or thick-walled containers of highly absorbing materials and even railcars. These types of high-energy X-Ray systems are very expensive and are typically only used for special applications.

3.4.1.4 X-Ray Sensors

An image is formed by exposing the package to X-Rays and detecting the transmitted or backscattered energy using a sensor or sensor array. The electronic signal produced by the sensor is processed, analyzed, and displayed as an image.

Sensors that employ scintillator materials in conjunction with photodiodes are among the most sensitive detectors. Incident radiation enters a scintillator crystal such as thallium-activated sodium iodide or cesium iodide. The energy produces photons of visible or near-visible ultraviolet light. These photons are converted to electronic signals by the photodiode, which is optically coupled to the crystal.

Other sensing materials that have been used in conjunction with detectors include phosphor (fluorescent) screens or strips. Phosphor screens are used to produce images of the entire package as in the case of fluoroscope imagers. Fluoroscopes are now used primarily in medical applications. For security purposes, they are typically used only in portable systems. Portable systems can be used to image the contents of abandoned packages; however, these systems are rather uncommon.

3.4.1.5 Imaging Systems

The operation of the imaging system is highly dependent on the exposure method and the sensor configuration. Fan beam imagers expose only a narrow strip of the package at one time. The transmitted energy impinges upon a linear array of X-Ray sensors. Typically, the linear array is a series of photodiodes that may be coupled with a strip of phosphor or scintillator material. The output of each photodiode in the array corresponds to a single pixel of the final image. The package is carried past the fan beam until the entire package has been scanned. Since the entire array is exposed at once, the output of the array is read serially and stored in the memory of the imaging system. After the entire package is scanned, the image is constructed and displayed.

Flying spot scanners can either be read by a linear array of point sensors or by a single strip sensor. The exact location of the spot on the package is known by the precise timing of the scanning system. The pixel information is stored in memory and displayed when the scan is complete. Radiation doses received by the package are very low in this method of scanning.

3.4.1.6 Image Processing System

Digital image processing allows a multitude of image enhancements that can assist the operator in seeing image detail. Edge enhancement sharpens or highlights any edge found in the image. This feature is helpful when the package contents are composed of many items that have similar X-Ray absorption characteristics. This situation normally produces an image of low contrast. Shapes that may be important can be lost in the clutter unless image processing helps to improve the contrast.

Intensity colorization is the simplest color enhancement that can be made to a system. In this process, colors are arbitrarily assigned to ranges of image intensity. Items that pass little energy may be colored dark blue, while items that pass most but not all of the X-Ray energy may be colored orange. Areas where almost all of the energy is passed may not be colored at all. The idea is that items that absorb little energy are organic and therefore may be explosives. The main problem with this feature is that the system cannot distinguish between very thin strong absorbers and thicker weak absorbers. For this reason, this approach has been largely abandoned. Systems that colorize material based on the atomic number (Z) of the material are a further enhancement but this technique requires at least dual-energy scanning.

Other image processing features that may be found on a system are zoom and pan and image negative display. Zoom and pan features allow an operator to enlarge a small area of the image that requires more in-depth investigation. Image negative display reverses the dark and light (or color) of the display because, for some operators, certain items stand out better in this display mode.

3.4.2 Operational Requirements

The conveyorized X-Ray baggage inspection system in airports is an example of an acceptable package search system in terms of size, weight, baggage throughput, and cost. Performance criteria for detection systems include the following important factors:

- High sensitivity

- High contrast and resolution

- Adequate scanning rates

- Good penetration

- Low dose to packages

- Discrimination between innocuous and threat materials

Many commercial X-Ray systems are available for screening packages. Effective X-Ray devices for detecting explosives and other contraband should meet the following criteria:

- The system should be capable of processing 10 packages per minute.

- The system may be single beam, single beam/dual energy, dual beam/single energy, or dual beam/dual energy and have one or multiple detectors.

- The operator should be able to discern whether a weapon, incendiary device, or explosive device is present in a typical briefcase, purse, and box.

Conveyor models can screen up to 25 packages per minute. System operation is continual and does not require a shutoff period. Operation is reliable up to 8,000 feet above mean sea level at temperatures from 0 to 50 degrees Celsius (32 to 122 degrees F) and at a relative humidity of less than 90 percent. External cooling is not required.

The image produced is typically displayed with sufficient resolution for the detection of 34 American wire gauge (AWG) wire; 14 to 20 shades of gray are detectable. An X-Ray tube voltage range of 65 to 200 kVp provides optimum contrast for the detection of high-density weapon steel and low-density plastic materials.

All units are certified safe for photographic materials (with the exception of CT scanners as discussed previously) and are in compliance with the U.S. Bureau of Health standards for cabinet X-Ray systems, which may be found in 21 CFR 1020.40, "Cabinet X-Ray Systems." These regulations limit radiation leakage to less than 0.5 millroentgen per hour (mR/h) to surroundings at 5 centimeters (2 inches) from cabinet surfaces. The units also comply with FAA standards found in 14 CFR 129.26, "Use of X-Ray Systems," and 14 CFR 108.17, "Use of X-Ray Systems."

3.4.3 X-Ray Techniques

Both transmission and backscatter effects are present when X-Rays interact with material. The magnitude of these interactions depends on the X-Ray energy, the density of the material, and the elemental composition (effective Z). The Z number of an element is the atomic number as found in a periodic table of the elements. For a compound, the effective Z number is derived from the Z numbers of the various elements and their proportion within the compound. Backscatter is useful for imaging low-Z materials, which appear bright in the backscatter image. Dual energy is also useful in discriminating between low-Z and high-Z materials. Various systems use a variety of techniques to obtain the X-Ray transmission rates at two levels of energy. A comparison of transmission rates at the two energies can be used to analyze the effective Z number. The low-Z materials can be identified and, through computer control, their images colored to ease operator identification. Regardless of the technique (backscatter or dual energy), low-Z material may require subsequent manual inspection because both explosive materials and innocuous organic materials are composed of low-Z elements.

3.4.3.1 Single-Beam, Single-Energy Transmission Systems

Whether a fluoroscope, a fan beam, or a flying spot, the single-beam, single-energy system is the simplest and least expensive system available. These systems provide black and white or intensity colorization images. These systems are useful mainly for examining a package for metallic items, such as firearms or knives, but are not usually useful for screening for materials like explosives or drugs. Effective screening of these low-Z number materials requires more sophisticated systems.

3.4.3.2 Dual-Energy Systems

The difference in technology between a dual-energy X-Ray system and a single-energy system is that material discrimination is achieved by comparing the ratio of low-energy X-Rays absorbed to high-energy X-Rays absorbed. Two basic systems can achieve dual-energy interrogation for low-Z materials. First, the output of standard X-Ray tubes is a fairly broad spectrum. Stacked linear arrays of photodiodes are used to create the image signal. The upper array absorbs many of the lower energy photons such that the X-Ray beam is "hardened" before it reaches the lower array. Since the average energy reaching the lower array is higher than the average energy seen by the upper array, dual-energy information can be obtained. Though the energy separation of these systems is relatively small, the scan can be faster and the overall system is simpler than systems employing the second method. The second method uses a single X-Ray generator operated sequentially with two voltages to provide the two energies; thus the energy separation can be as large as desired.

The resulting image (refer to Figure 9) is displayed on a monitor for visual identification. The software characterizes and identifies the various materials by image shape, and artificial colors are assigned to various Z number values.

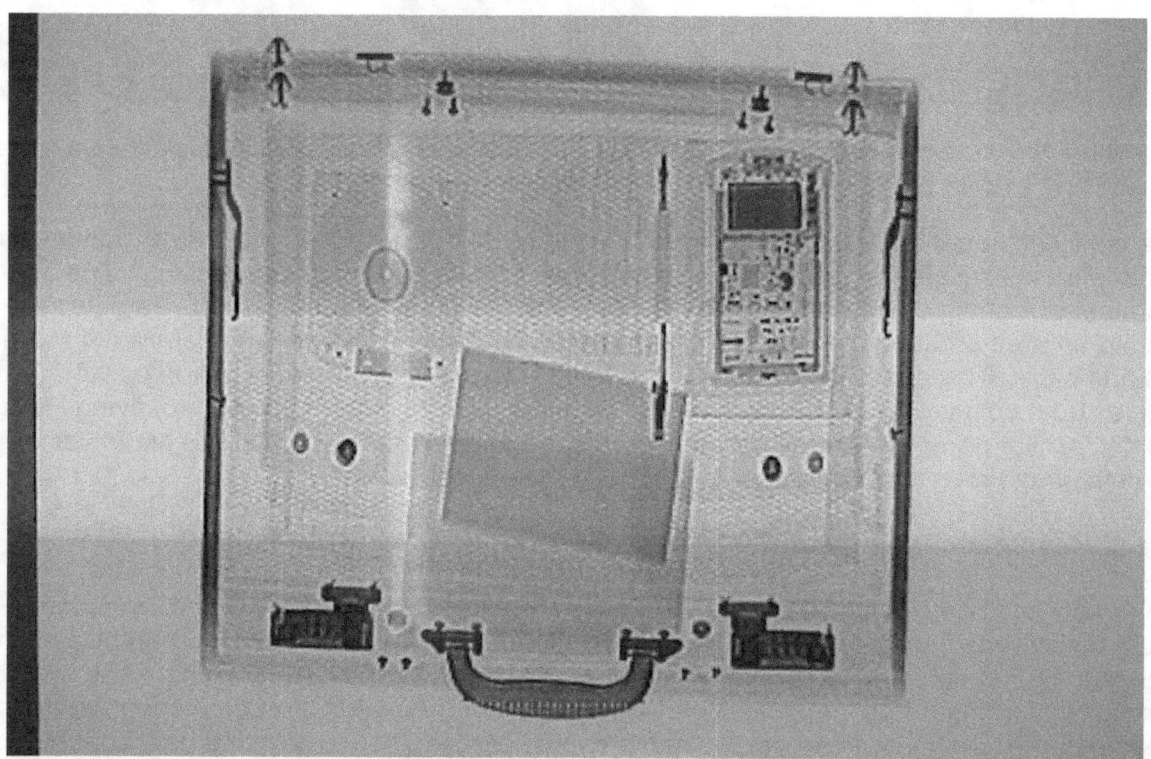

Figure 9: Dual-energy image of a briefcase containing polymer used to simulate explosive material. Note that the rectangle of darker orange (polymer) is easily seen in the cluttered briefcase. The lighter orange partially covering the polymer is a paperback book.

3.4.3.3 Backscatter Techniques

Low-Z materials are more efficient at backscattering X-Rays than high-Z materials, primarily through Compton backscatter and, to a lesser degree, through Raleigh backscatter. For this reason, backscatter imaging systems are designed to provide both a standard transmission image and a second image based on the energy backscattered from the package (Figure 10). In a transmission image, the low-Z materials appear ghostly and low contrast, while metallic objects stand out. This makes the transmission image better for screening for metallic weapons. In the backscatter image, the low-Z materials appear bright and high contrast, while the metallic items are low contrast. This makes the backscatter image better at screening for low-Z materials, such as explosives.

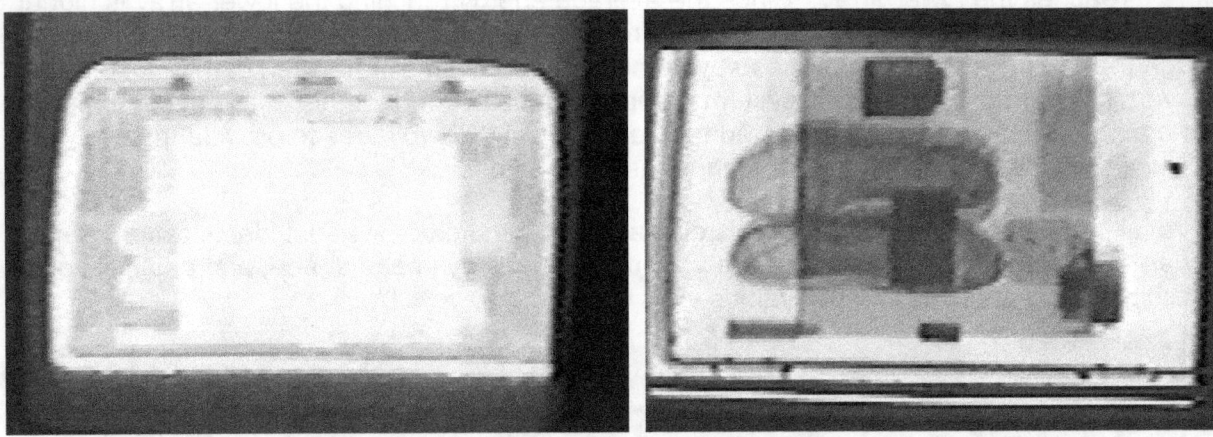

Figure 10: Backscatter and transmission images of the same package. The image on the left is the backscatter image and the image on the right is the transmission image.

Because flying spot backscatter imaging techniques result in such low dose rates, a system can be made that is suitable for screening personnel for objects hidden under clothing. The personnel screening devices deliver about 2.5 microrem per side scanned; thus, a person entering an area protected by one of these systems would receive about 5 microrem if screened front and back. This can be compared to the dose rate that a commercial airline passenger receives from cosmic radiation, which is approximately 300 microrem per hour of flying. More recently, similar personnel imaging systems have become available that utilize nonionizing millimeter-wave radiation.

3.4.4 X-Ray Scanning Procedures

All hand-carried items (such as purses, briefcases, packages, lunch boxes, and overcoats) should be screened separately from the personnel carrying them. All hand-carried or personal items that cannot be screened properly because of composition or contents should be physically searched or excluded from controlled areas. As a person proceeds through a contraband screening area, his or her hand-carried items should proceed separately and be inspected for contraband at the same time the individual is being inspected. All delivered packages and bulk items should also be screened before being taken into controlled areas.

Under optimal conditions, two individuals who are trained and qualified in all aspects of search operations and should be stationed at an X-Ray scanner. One should assist, control and observe personnel processing through the search train while the other operates the X-Ray scanner. The length of time that an individual operates X-ray scanning equipment should be limited to ensure that the individual's ability to recognize contraband in the display image is not compromised due to fatigue associated with lengthy periods of operation. Optimally, if two trained and qualified search personnel are performing search operations, the two individuals should switch positions approximately every 30 minutes.

Each package should be placed in the manufacturer's recommended orientation in the center of the conveyor and passed through the X-Ray scanner. If the image reveals any shape, coloration, or other feature that may be contraband, procedures should be defined to positively identify the item in question. Procedures for positive identification may include the following:

(1) Pass the package through the X-Ray scanner a second time.

(2) Reverse the package (i.e., flip it over) and pass it through the X-Ray scanner a third time.

(3) Physically search and inspect the package.

(4) Call for the additional assistance of another member of the security organization to accompany the person and package to a holding area, if necessary.

If any of these steps resolves the issue, the subsequent steps do not need to be taken. If an individual refuses to comply with screening procedures, or if a weapon, explosive, or other unauthorized item is found, entry should be denied and the person should be escorted to a holding area for appropriate action. If material of a suspicious or unknown nature is found, entry should be denied until the situation can be resolved by the appropriate level of security supervision.

3.4.5 Calibration

Typically, commercial X-Ray imaging systems have no user calibration adjustments. Facilities typically can only conduct a performance test with one of the test objects described below and then obtain field service from the manufacturer if and when the system performance starts to degrade.

3.4.6 Operability Testing

At the beginning of each security duty shift, the system operator should pass a test object through the X-Ray device to ascertain that the device is functioning as intended. The test object could be one of the two American Standards for Testing and Materials (ASTM) test objects described in detail in the next section. Using the ASTM step wedge, an adequate operability test is to note that an image of the wedge is produced and that the 30 AWG wire can be seen beneath the thickest step. Any catastrophic failure of the X-Ray device would be immediately apparent as a result of the failure of the device to produce an image of an item being screened.

3.4.7 Test Objects and Performance Testing

The section that follows describes performance tests that can be carried out with two different test objects developed by ASTM. The step wedge is an older and less sophisticated test object, so the use of the new test object for performance testing is preferred. Simple tests with an explosive simulant test object are also discussed. Performance testing of X-Ray imaging systems should occur once every 7 days.

3.4.7.1 Performance Test Using the ASTM Step Wedge

ASTM Designation F792-88, "Standard Practice for Design and Use of Ionizing Radiation Equipment for the Detection of Items Prohibited in Controlled Access Areas," describes the ASTM step wedge and its use. This device allows test personnel to determine the sharpness (often referred to as "resolution" in X-Ray manufacturers' literature) of the X-Ray image, the penetration capability of the system, and the contrast of the system. The wedge is constructed mainly of aluminum, milled into 10 steps of increasing thickness. The first step is 0.069 inches, with each step increasing by 0.069 inches. Affixed to the bottom of the wedge is a series of wires arranged in a sinusoidal pattern. The wires range from 22 to 30 AWG in steps of two AWG, for a total of five wires. The image of the wedge is read by first determining that all 10 steps of the wedge are penetrated and that all 10 steps are clearly visible. This is a measurement of the image contrast and penetration. The next evaluation step is to determine the thinnest gauge of wire that can be imaged beneath each of the 10 steps. A modern X-Ray system should be capable of easily imaging all 10 steps clearly and of imaging the 30 AWG wire under the thickest step. See Figure 11 for a representation of the ASTM step wedge.

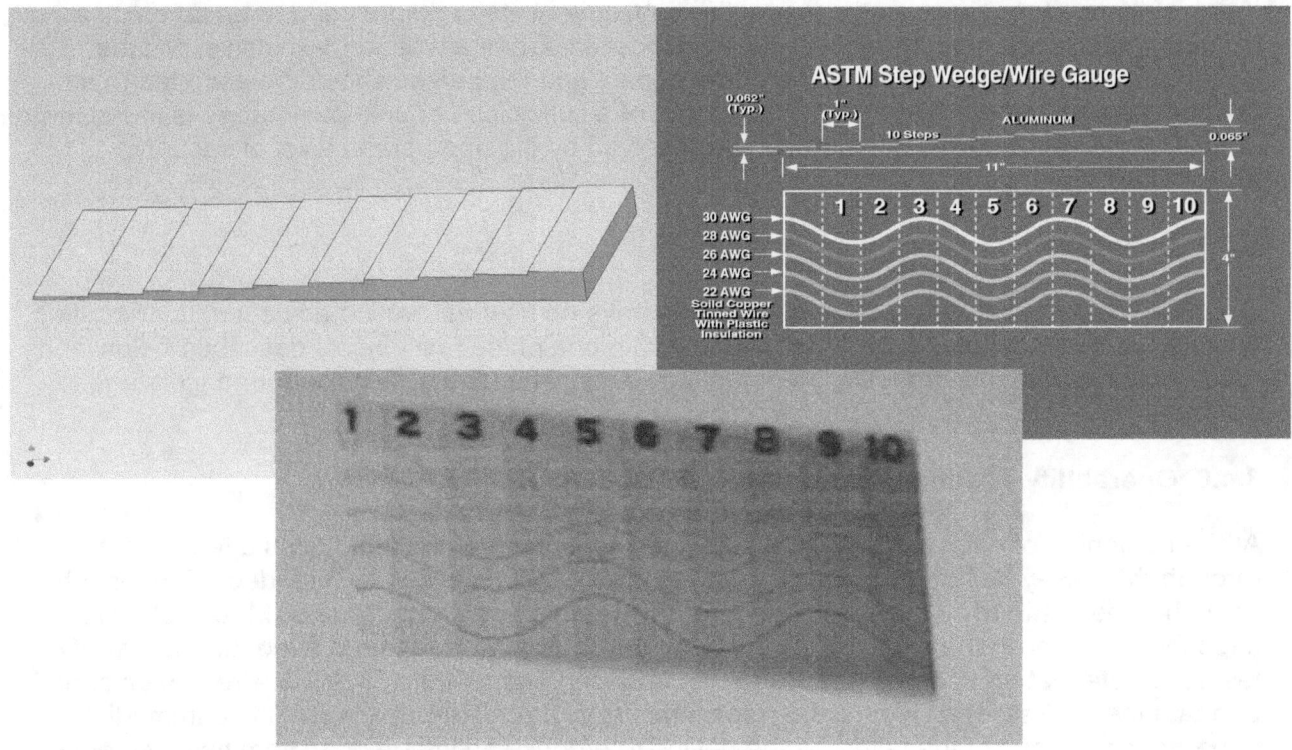

Figure 11: The ASTM step wedge. The upper left shows a rendering of the visual appearance of the aluminum wedge; while the upper right drawing shows the wire arrangement and gives the dimensions. The bottom graphic shows an image of the wedge displayed by an X-Ray imaging system.

This test object is still in use but has been largely replaced by the newer test object described below, which measures more aspects of system performance and adds tests to measure calibration of the more sophisticated enhancements, such as dual-energy analysis. It is thus recommended that this step wedge be used primarily for operability testing and that performance testing be done with the new ASTM test object described below.

3.4.7.2 Performance Test Using the New ASTM Test Object

ASTM F792-01, "Standard Practice for Evaluating the Imaging Performance of Security X-Ray Systems," updated in 2001, defines the new test object. The new ASTM X-Ray test object (shown conceptually in Figure 12) is to be used throughout the test, as described in the "Standard Practice," to determine the applicable performance levels of a system. This test object is sufficiently complex that performing all possible tests with the object constitutes an acceptable performance test for an imaging system. However, a subset of tests similar to those performed with the step wedge could still be performed as an operability test.

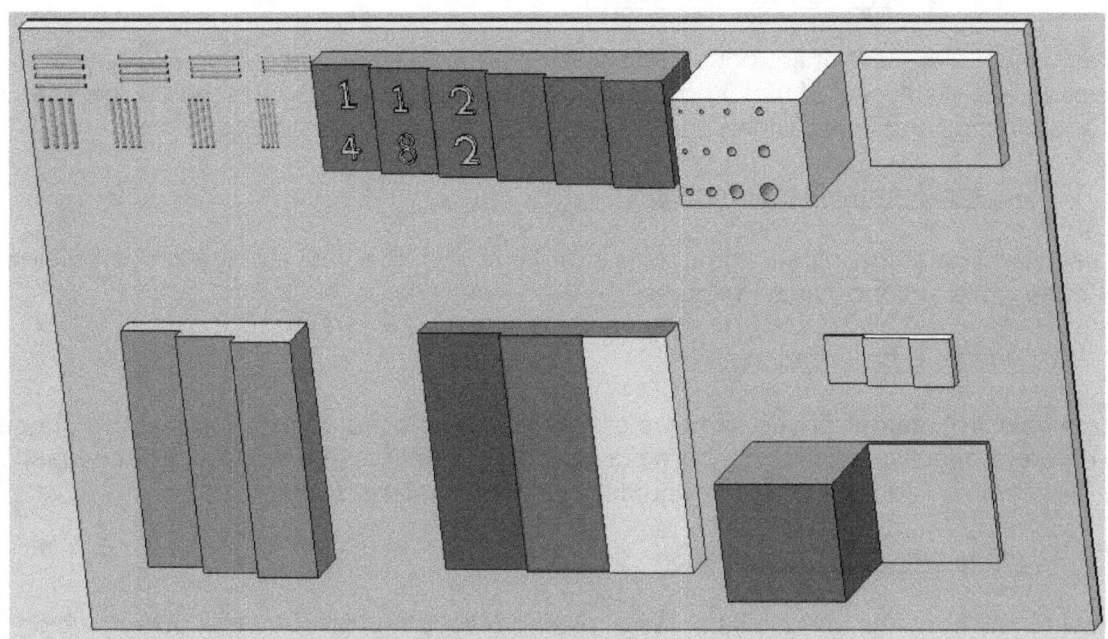

Figure 12: New ASTM test object, described in F792-01

The test object was developed to assess the image quality of an X-Ray-based screening system in the following nine distinct areas:

3.4.7.2.1 Wire Display

The test object contains a range of sizes of wire gauges.

3.4.7.2.2 Useful Penetration

The varying gauge of the wires embedded in the object and the thickness of the aluminum provide sufficient range to characterize the system's useful penetration.

3.4.7.2.3 Spatial Resolution

The test object contains a set of narrowly spaced wires with a particular gauge.

3.4.7.2.4 Simple Penetration

The test object contains several digits (0 through 9) made from lead on top of a steel object that varies in thickness.

3.4.7.2.5 Thin Organic Imaging

The test object contains plastic of various thicknesses.

3.4.7.2.6 Image Quality Indicators Sensitivity

The test object has a set of flat-bottomed holes drilled into steel and plastic samples that vary in thickness. The diameters of these holes, the depths of these holes, and the thicknesses of the steel and plastic samples provide various ranges.

3.4.7.2.7 Organic/Inorganic Differentiation

The test object contains a steel and plastic sample to test how well the system can differentiate between organic and inorganic materials.

3.4.7.2.8 Organic Differentiation

The test object contains various samples of different plastic materials. The plastics chosen have different effective atomic numbers but nominally identical attenuation. These plastics are used to resemble various foodstuffs, liquids, and other materials.

3.4.7.2.9 Useful Organic Differentiation

The test object contains various samples of plastic placed on top of a steel object, which varies in thickness.

The following form (Figure 13) is suggested to record the data collected using the new ASTM test object. ASTM F792-01 provides details on the use of this form.

ASTM X-RAY TEST OBJECT LOG SHEET

DATE——————— TIME ——————— OPERATOR ——————————————

X RAY MAKE ——————— MODEL ——————— SERIAL # ——————— SOFTWARE VERSION ———————

MONITOR MAKE——————————— MODEL ——————— SERIAL # ———————

POSITION ON BELT (Left, Middle, Right)——————— ORIENTATION (Vertical, Horizontal) ———————

TEST	IMAGING OPTIONS USED
1 Wire Resolution	
2 Useful Penetration	
3 Spatial Resolution	
4 Simple Penetration	
5 Thin Organic Imaging	
6 IQI Sensitivity Test	
7 Organic/Inorganic Differentiation	
8 Organic Differentiation	
9 Useful Organic Differentiation	

TEST 3

2.0   1.6   1.3   1.0

TEST 1

40   36   32   30   24

TEST 1
TEST 2

9 5   15 9   22 2

TEST 4

1   4
1   8
2   2
2   6
3   0
3   4

TEST 6

1 2 3 4   4T 2T 1T
1 2 3 4 5

0 48   0 32   0 16

TEST 9   TEST 8

TEST 7

TEST 5

1
2
3

TEST 7

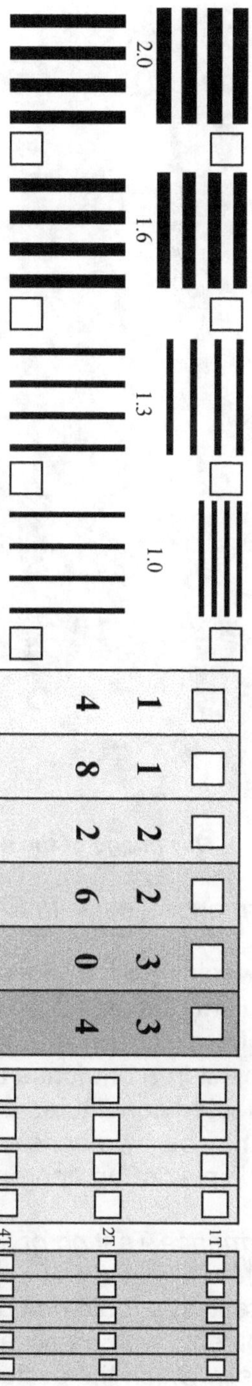

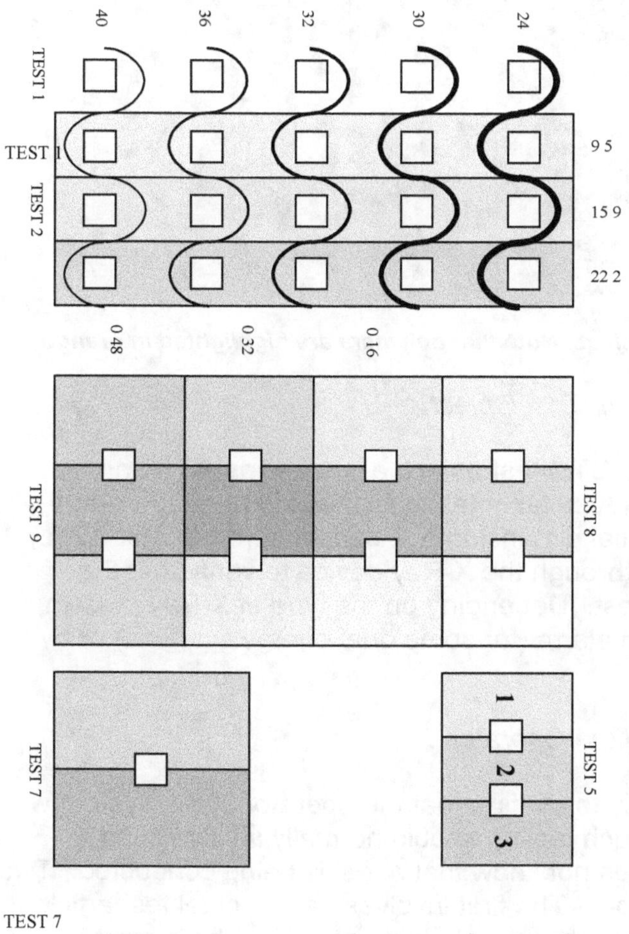

Figure 13: ASTM X-Ray test object log sheet

3-27

Figure 14 shows an X-Ray image of the test object as recorded when placed inside of a suitcase.

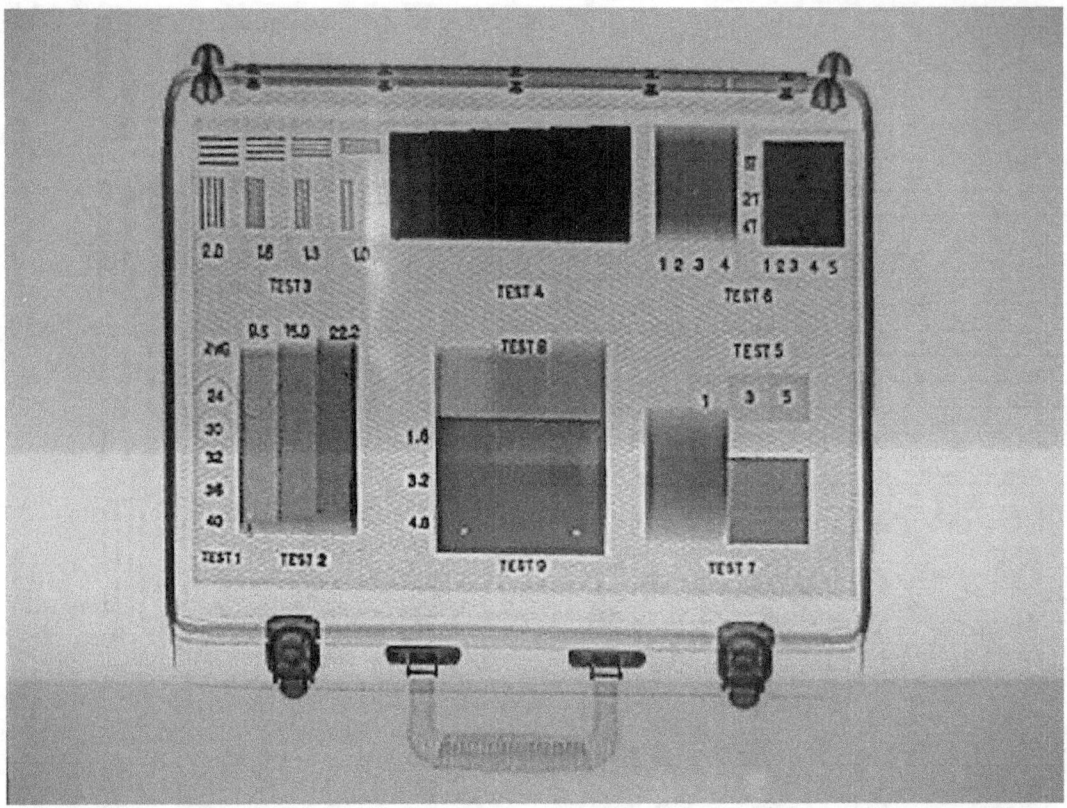

Figure 14: X-Ray image of the new ASTM test object. Note that polymers are highlighted in orange.

3.4.7.3 Performance Test with Explosive Simulant Test Object

In addition to weekly performance tests with an ASTM test object, a weekly test with an explosive simulant test object is recommended. A suitable test object would be a 200-gram (7 oz.) piece of acetal resin with a minimum spatial dimension of 2 centimeters (approx. ¾ in.). Placing this object in a briefcase and passing it through the X-Ray device to verify that the simulant is detected constitutes an acceptable test. Depending on the type of X-Ray system, the detection may be by means of an automated alarm (for some dual energy systems) or by means of the display of an appropriate image.

3.4.7.4 Performance Testing of the System and the Operator

Performance testing should be conducted weekly to confirm that the operator of the system will detect threat objects with an acceptable rate. Such testing should normally involve blind challenge tests in which the system operator does not know that a test is being conducted. Two possible approaches can be used to test operators. The first involves using actual test articles placed inside packages, while the second involves the use of electronic TIP. The advantages of using TIP instead of actual items include the following:

- Entering personnel will not be aware that their package was used in performance testing.

- An operator will not know that he or she is being tested because the system uses a random process to project test images. (It is possible to perform blind tests with actual threat objects as well, but it is not as easy.)

- A larger assortment of threat items can be used than is practical when using actual items. For example, the extensive TIP libraries used to train X-Ray screeners in the United Kingdom utilize 500 images, divided into the three categories of improvised explosive devices, guns/knives, and liquids/gels.

- Actual threat items are not present in the screening area.

- Training time, effort, and cost will probably be reduced.

Testing by either method usually results in similar data. The number of challenge tests is recorded, along with the number of times the operator detected the presence of the threat. Analysis of performance is calculated using the same methodology as for other detection systems. A running record of performance should be maintained for all system operators. If a particular operator shows poor performance compared to his or her peers in detecting threat items or images of threat items, remedial training of that operator may be necessary. Operators should undergo an initial certification test and be recertified annually. No well-defined detection efficiency can be quoted as a minimum requirement, since this depends on a number of factors, such as the amount of clutter in a screened item and the distinctiveness of the image projected. However, the rate of detection should be close to 100 percent for certain easy-to-detect images, such as the image of a gun projected onto the top middle of a bag's contents. Table 4 summarizes the types of operability and performance tests to be performed with an X-Ray inspection system.

3.5 Computed Tomography

This section provides guidance for the testing of CT systems at protected area and material access area entrances. *Note: CT systems are designed for automated detection of explosive-like materials and not for the imaging of other threat items such as guns.*

3.5.1 Operational Principles

CT systems for the detection of bulk explosives are scanning X-Ray systems originally based on medical equipment. The system scans a package using a spinning gantry that spins around the package, exposing it from all sides. Once a series of scans is complete, the system reconstructs images of the interior of the package in three dimensions. The system can determine the X-Ray mass absorption coefficient of any material within the package and generate an alarm whenever a material above a mass threshold and with a mass absorption coefficient within the range for explosives is present.

Table 4: Summary of recommended testing for X-Ray inspection systems

Test	Test Object	Test Frequency
Operability Test	ASTM step wedge (verify only that the 30 AWG wire can be seen through the thickest step)	Start of security duty shift
Performance Test (system only)	ASTM new test object (note and record all image details appropriate for that object)	Weekly
Explosives Simulant Performance Test (system only)	Acetal resin 200 grams (7 oz.) (verify that an appropriate image or automated alarm results)	Weekly
Performance Test (system and operator)	TIP or blind challenges with actual or mock threat items	Initial training for new operator and periodic testing thereafter. Maintain running record of operator performance. Remedial training may be necessary if an operator exhibits continued poor performance.

3.5.2 Performance Standards for CT Scanning

The very high cost of CT systems currently limits their use for the detection of bulk explosives. At present, the greatest user of CT for security is the Transportation Security Administration (TSA). Only the FAA—now part of the TSA—has established CT performance standards, although these standards are not published. Until CT is more widely used, it is unlikely that industry performance standards will be established. This document discusses how performance can be measured to confirm continued, consistent performance over time after installation.

3.5.3 Test Objects for CT Scanning

The next sections provide guidance for the selection of test objects that can be used to test CT systems for detection performance.

3.5.3.1 Simulants

Because the use of real bulk explosives for routine testing of CT systems is too risky, using substitutes for explosives is desirable. Often the materials used as simulants are polymers, since the linear absorption coefficients for some polymers lie within or very near the range of linear absorption coefficients typical of explosives.

3.5.3.2 Mass Absorption Coefficients

Almost all common explosive materials are composed of light elements. Commercial explosives are composed of carbon, oxygen, hydrogen, and nitrogen. Each element has a mass absorption coefficient that can be found in tables in many reference sources. The mass absorption coefficient for a given element is wavelength dependent, and the coefficient decreases as the wavelength becomes shorter (i.e., as the X-Ray energy increases). This is expected since X-Rays become more penetrating as the wavelength shortens (less energy is absorbed). While the mass absorption coefficients for elements can be readily found in reference books, the mass absorption coefficients for compounds and homogeneous mixtures of compounds must be calculated. To perform this calculation, sum the products of the mass absorption coefficient of each element multiplied by the mass fraction of that element in the compound or mixture. For instance, if carbon constitutes 1/16[h] of the mass of a compound, multiply 0.0625 (1/16) by the mass absorption coefficient for carbon at the given wavelength. This step is repeated for each element in the compound, and all the resulting products are summed. More useful for simulant analysis is the linear absorption coefficient. The linear absorption coefficient is simply the mass absorption coefficient multiplied by the mass density of the element, compound, or mixture.

3.5.3.3 Linear Absorption Coefficients

A CT system alerts on a material if the linear absorption coefficient of that material lies in the rather narrow range of linear absorption coefficients typical of explosives. Table 5 lists the mass absorption coefficients for common elements found in polymers and explosives at a wavelength (expressed in reciprocal centimeters/cm^{-1}) of 2.29 angstroms (Å) (1 Å = 1.0 x 10^{-10} meter).

Table 5: Mass absorption coefficients for common elements in polymers and explosives at 2.29 Å (in cm^2/g)

Element	Mass Absorption Coefficient
Carbon	14.20
Oxygen	39.40
Hydrogen	0.50
Nitrogen	23.10
Fluorine	50.30

The following graph (Figure 15) shows the linear absorption coefficients for several common explosives and several common polymers.

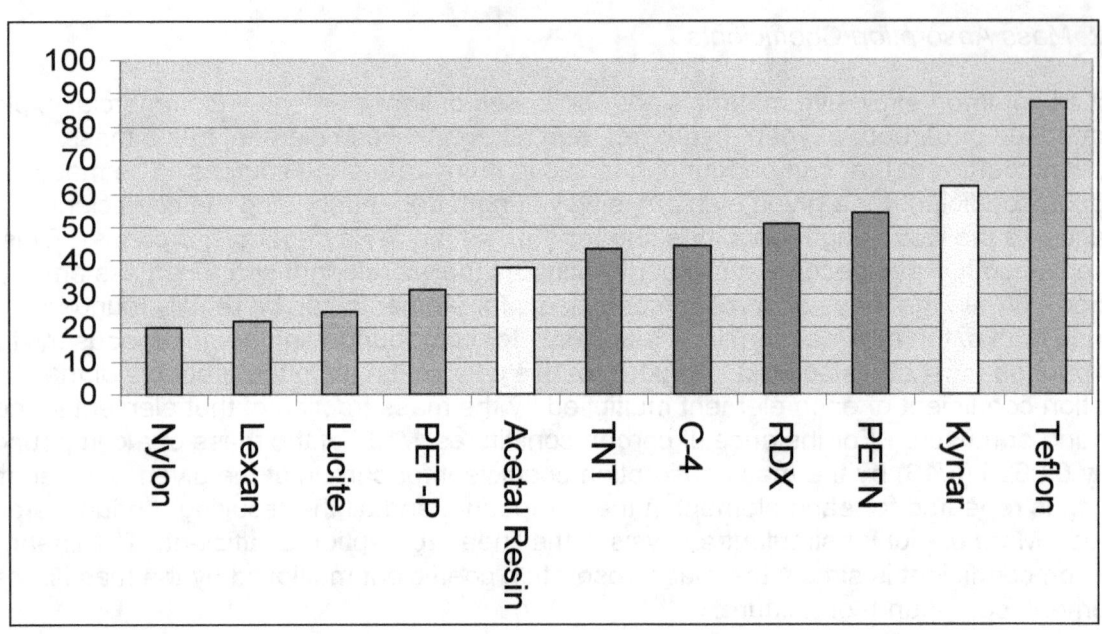

Figure 15: Selected X-Ray linear absorption coefficients at 2.29 Å

The explosives lie in a relatively narrow range of linear absorption coefficients between 43.32 cm^{-1} (trinitrotoluene (TNT)) and 54.13 cm^{-1} (pentaerythritol tetranitrate PETN). The lighter polymers shown in blue will likely not be suitable simulants for explosives, with the possible exception of PET-P (pentaerythritol tetranitrate Polyester, a common polyester used to make soda and water bottles). The acetal resin (common name Delrin®) will in most cases act as a reasonable simulant for explosives. Likewise, Kynar® may serve as a suitable simulant while Teflon® will not. Teflon® has a high linear absorption coefficient resulting from the presence of fluorine atoms in this material. Note that C4 has a lower linear absorption coefficient than RDX (Research Department Explosive, chemical name: cyclotrimethylenetrinitramine), even though the explosive constituent of C4 is RDX. This is caused by the presence of the lower density plasticizers used to make RDX, which is a powder, into the plastic explosive C4. In the same way, a homogenous mixture of two polymers, in which one has a linear absorption coefficient lower than explosives and the other has a linear absorption coefficient higher than explosives, can have a combined mass absorption coefficient in the middle of the range for common explosives. Custom blends of these polymers could be fabricated to exactly match any desired linear absorption coefficient, although this is not typically done. Usually, a few commercial plastics have a linear absorption coefficient that is close enough for testing purposes.

3.5.4 Calibration of CT Systems

CT systems may require factory calibration periodically. There are typically no user adjustments for calibration purposes. Users should consult with the manufacturer if system performance degrades or there is any other indication that the system may not be functioning properly.

3.5.5 Operability Testing

Operability testing should be conducted at the start of each security duty shift to confirm that the system will reliably detect the simulant material. A standard simulant test object can be placed in a briefcase and passed through the system and an alarm should result if the system is operating correctly. A reasonable test object is one composed of acetal resin, with a mass of 200 grams (7 oz.) and a minimum dimension of 2 centimeters (approx. ¾ in.) (i.e., all three spatial dimensions of the object are 2 centimeters (approx. ¾ in.) or greater). If the system fails to detect a test object that it has been reliably detecting since installation, the user should consider contacting the manufacturer to arrange for calibration or repair. Consultation with the manufacturer is also recommended to make certain that the material and dimensions selected for the standard simulant test object are appropriate.

3.5.6 Performance Testing

Performance testing should be conducted weekly to monitor various aspects of system performance. Tests can be designed to do the following:

- Vary the amount of plastic explosive simulant and compare the alarm rate for simulant test objects of different shapes and sizes. Acetal resin test objects with masses from 200 to 1,000 grams (7 to 35 oz.) can be utilized as long as all objects have a minimum dimension of at least 2 centimeters (approx. ¾ in.). Different shapes could include a cube, various rectangular solids, and a cylinder.

- Vary the amount of clutter in a briefcase or other container that contains the standard simulant test object.

- Screen briefcases or other containers that contain objects which may produce false alarms. Monitoring and recording all alarms during screening over an extended time period will give an accurate assessment of the CT system's false alarm rate in the real world setting where it is utilized. By noting what items can produce false alarms, a test kit of such items could be developed.

Table 6: Summary of recommended testing for CT systems

Test	Test Object	Test Frequency
Operability Test	Acetal resin 200 grams (7 oz.) in briefcase	Start of security duty shift
Performance Test (vary mass and shape of simulant)	Acetal resin from 200 grams to 1,000 grams (7 to 35 oz.) in briefcase. Use different shapes* (e.g., square block, rectangular block, or cylinder)	weekly
Performance Test (vary briefcase clutter)	As in operability test, but increase number of additional items in briefcase	weekly
Performance Test (false alarms)	Actual items screened in real world	Maintain continuous records of items that produce alarms and percentage of screened items that alarm
*with minimum dimension of 2 centimeters (approx. ¾ in.)		

3.6 Trace Explosive Detection Equipment

3.6.1 Test Objects for Trace Explosive Detection Equipment

A variety of test objects can be used to test trace explosive detection equipment, whether the technology being utilized is IMS or another trace technique. Test objects fall into two categories:

- (1)   trace explosive material

- (2)   bulk explosives

Since safety regulations often prohibit the use of bulk explosives, the focus here will be on testing trace detection equipment using only trace explosive material.

3.6.1.1 Creating Test Objects for Trace Explosive Detection Equipment

Most tests will utilize either particle contamination or explosive material deposited from a standard solution.[2] Particle contamination can be produced easily by directly or indirectly contacting sampling swabs or clothing (for personnel portals) with explosive material. For example, the person preparing a sample can touch bulk explosive material or even a container that holds such material and then place a fingerprint on a sampling swab. Creating test samples in this manner may be done at a storage bunker or some other remote location rather than at the location where the equipment is deployed, at least if low-vapor-pressure explosives, such as TNT, RDX, and PETN, are being utilized. Test samples prepared in this manner contain unknown quantities of trace explosives, so this will result in a qualitative (present/not present) rather than a quantitative ("X" amount) test. As stated previously, most detection equipment manufacturers offer test kits or specific test objects that challenge the detection capabilities of the detection system through the entire range of threat objects for which the system is designed to detect. In many cases the use of test kits offered by the detection equipment manufacturer is the easiest, most efficient, and safest solution for detection system testing.

3.6.1.2 Detectable Amounts of Explosive Materials

Commercial IMS systems can detect nanograms or even picograms of explosive material (1 nanogram = 10^{-9} gram; 1 picogram = 10^{-12} gram), and most detectors can easily detect a deposited fingerprint if the sample is presented to the detector in the correct manner. In fact, the danger of saturating the detector and forcing a protracted period of cleanup in greater than the danger of not being able to detect a fingerprint sample.

3.6.1.3 Creating Sample Swabs with Known Amounts of Explosive Material

A sometimes preferable test object is a sample swab onto which a known amount of trace explosive material has been deposited from a standard solution. Standard solutions of

---

[2]   In principle, explosive vapor can be used to challenge trace detection equipment. However, explosive vapor generators tend to be experimentally complex and development of them has been limited, so this report does not address vapor generators.

explosives in solvents, such as methanol or acetonitrile, are available commercially, and portions of these solutions can be diluted to produce daughter solutions at any desired concentration. A clean syringe is used to collect a small amount of solution and deposit it onto the swab or other sampling medium for presentation to the instrument. For example, a 1-microliter syringe can be used to collect 1 microliter of solution that contains 1 nanogram per microliter of TNT; when this solution drop is deposited onto a swab and a few seconds have elapsed to allow the solvent to evaporate, the result is a swab containing 1 nanogram of TNT. The swab can then be placed in the analysis port of a benchtop IMS detector and analyzed. Not only should detection of the TNT result, but the magnitude of the TNT signal will provide an approximate or semiquantitative calibration of the detector response to TNT. It is only semiquantitative because IMS is inherently not well suited to quantitative analysis, but even a semiquantitative result has considerable value.

3.6.1.4 Multiple Tests for Trace Explosive Detection Equipment

More information on the explosive detector functionality can be obtained if the detector is tested with several different TNT masses or several times with the same mass. A good compromise between time expended and information obtained is to test the detector one time each with two appropriately selected TNT masses that are approximately an order of magnitude apart, for example 0.5 nanogram and 5 nanograms. The equipment vendor should be able to provide information on what masses might be appropriate, but both masses need to be greater than the smallest mass the system can detect and smaller than the mass at which the detector signal saturates. Similar tests can also be conducted with other explosives of interest if standard solutions are available.

3.6.1.5 Running "Blanks" for Trace Explosive Detection Equipment

A key point with any piece of trace explosive detection equipment is that blank (clean) samples must be run after any detection of explosive material has been made, regardless of what sort of test or screening was being performed. If running a blank sample results in continued detection of explosives, the detector may have become contaminated and more blanks will need to be run until no more trace explosive is detected. Running of blanks after an explosive alarm should always continue until at least two consecutive blanks result in no detection of explosives. The following describes the process of running a blank for each type of detector:

- For a benchtop trace detector, insert a clean sample swab into the analysis port and analyze it.

- For a hand-held detector, run the system while drawing only room air into the sampling inlet.

- For a personnel portal, operate the portal with no one standing in it.

If explosive alarms continue to result for numerous blank runs, the equipment contains persistent contamination and the user's manual should be consulted to determine how to clean the unit before resuming testing or screening.

3.6.2 Calibration of Trace Explosive Detection Equipment

Calibration of trace explosive detection equipment should be performed at the start of each security duty shift and at the end of the security duty shift. Calibration should also be performed if the equipment performance appears to deteriorate, if any instrumental settings are changed, or if the equipment is moved to a new location. Bear in mind that changing environmental conditions, such as temperature, pressure, and humidity, can lead to deteriorating performance that may require recalibration.

Calibration of trace explosive detection equipment normally involves challenging the system with some type of trace material provided by the manufacturer. For example, one benchtop detection system provides a "lipstick-like" material that contains several common high explosives. The operator simply touches a sampling swab lightly with the "lipstick" and then analyzes the swab. A correctly functioning system will then detect the various explosives contained in the "lipstick." Other types of trace detection equipment, such as some hand-held and portal detectors, have automatic calibration buttons that, when pushed, trigger the release of trace material into the system automatically. The Transportation Security Laboratory (TSL) in Atlantic City, NJ (now part of the U.S. Department of Homeland Security) has developed a test kit composed of two solutions that contain RDX and PETN, respectively, which is referred to as the TSL Liquid Quality Control Kit. This test kit can be obtained from the TSL and may be of interest to some field users of trace detection equipment.

3.6.3 Operability Testing of Trace Explosive Detection Equipment

Operability testing of trace detection equipment can utilize tests with TNT samples from standard solutions as discussed previously. Depending on the time available and the amount of information desired, testing can be done with more or fewer replicate runs, types of explosives, and different masses. Testing a benchtop IMS detector with two different TNT masses and two different RDX masses is an adequate operability test. Similar testing is appropriate for hand-held detectors and walkthrough portal systems, though for these systems the presentation of the sample to the system may be more complicated than simply placing the sampling swab in the instrument's analysis port. For a hand-held detector, the sample may need to be agitated in front of the instrument's sampling inlet since most hand-held detectors are designed more for vapor detection than for analysis of particulates. For a personnel portal, the swab with trace explosive material could be either taped to someone's clothing before that person enters the portal or agitated in front of the portal's air collection inlet. Table 7 provides an example of a set of tests that could be used for the operability testing of a benchtop IMS system.

Table 7: Sample operability test for a benchtop trace detector based on IMS detection

Test Object	Detector Response
TNT mass x on sample swab	a counts
TNT mass 10x on sample swab	b counts
RDX mass y on sample swab	c counts
RDX mass 10y on sample swab	d counts

3.6.4 Performance Testing of Trace Explosive Detection Equipment

Typically, a wide variety of tests will be involved in performance testing of trace explosives detection equipment. Some tests need to be "blind," and deposition from a standard solution onto a readily visible sampling swab will not be appropriate for these tests. Rather, personnel or other items to be screened should be deliberately contaminated with nonvisible trace explosive material without the knowledge of the screeners. Some examples of appropriate tests would include the following:

- For a benchtop trace detector used to analyze swipes obtained from laptop computers or other hand-carried items, deliberately place fingerprints containing explosive contamination or trace material from a standard solution on these objects.

- For a benchtop trace detector used to obtain swipes from the hands of entering personnel, send people through the checkpoint with deliberately contaminated hands.

- If a walkthrough personnel portal is used, send people through whose clothing has been deliberately contaminated.

Care needs to be taken not to overdo the amounts of trace contamination to avoid instrument shutdown resulting from significant internal contamination that will necessitate lengthy cleaning procedures. Consultation with the equipment vendor and some initial experimentation may be required to determine approximately what the correct amounts of contamination are (e.g., one fingerprint, 10 fingerprints). Some key test variables to investigate include the following:

- Several different types of explosives should be investigated. For example, TNT, RDX, PETN, and NG (nitroglycerin) might form a typical performance test set for an IMS detector.

- The location of the contamination on a person or test object can be altered. This is especially important in the case of personnel portals, where separate test runs might be performed with contamination on both the front and back of a person and above and below the waist.

- The equipment operator should be varied. Some testing should be performed with each of the different personnel who have been trained to operate the equipment.

- The amounts of explosive contamination might be varied; however, since IMS is at best semiquantitative and since many test samples will not be easily quantified, this may be of limited utility. Furthermore, as stated previously, too much contamination can overwhelm the detector and require a long cleanup process.

Table 8 summarizes these test variables. Depending on site needs, performance testing could be conducted over a relatively short time period or over the course of a period as long as 1 year.

Table 8: Experiments that may form part of a performance test for a trace detection system

Test Variable	Possible Choices or Values
Explosive Type	TNT, RDX, PETN, NG, perhaps others. RDX could be in the form of C4 or Semtex; PETN in Detasheet or Semtex; NG in dynamite, etc.
Explosive Location	For personnel portals, test front vs. back and near feet vs. waist vs. near shoulders. For benchtop used to swipe hand-carried items, use test objects with trace contamination at different locations.
Detection System Operator	Operator A, Operator B, Operator C, etc. (Include all individuals trained to operate the equipment.)
Trace Explosive Mass	Mass X, Mass 10X, Mass 100X (Smaller mass increments may be appropriate.)

3.6.5 False and Nuisance Alarms for Trace Explosive Detection Equipment

Another important aspect of performance testing is determination of the approximate rate of false and nuisance alarms.

False alarms are alarms that are not caused by explosive material; nuisance alarms are alarms resulting from real trace explosive material (such as medicinal nitroglycerin) when no threat object (such as a bomb) is present. Because it is often impossible to determine whether a specific alarm is a false alarm or a nuisance alarm, the most relevant number is the combined rate of false and nuisance alarms, which may be referred to as the false/nuisance alarm rate.

3.6.5.1 Nuisance Alarms

Nuisance alarms can result from a variety of sources, including people who have handled an explosive as part of their jobs or otherwise worked in areas where explosives are used, people who handle firearms, people who work with certain types of fertilizer, and heart patients who take nitroglycerin tablets.

3.6.5.2 False Alarms

False alarms may result from equipment malfunctions or unknown causes. Alarms may also occur in which a trace detector mistakes a nonexplosive compound for an explosive that has certain similar chemical properties. Although such alarms are in some cases referred to as nuisance alarms, they are more properly labeled false alarms because the detector has alarmed for an explosive in the absence of any trace explosive material.

Determination of the false/nuisance alarm rate can be made over an extended time period simply by recording all screening results and dividing the total number of alarms by the total number of people or objects screened (assuming no explosives have actually been discovered). Unless a site has a very high level of explosive background contamination, this rate should not be higher than a few percent of all objects screened.

3.6.6 Alarm Resolution

Another important topic is alarm resolution, or, specifically, the procedures that should be followed if a person or object being screened produces an explosive alarm. Facilities can develop decision trees to assist security personnel who conduct screening to understand the process of alarm resolution. Figure 16, depicts a sample decision tree.

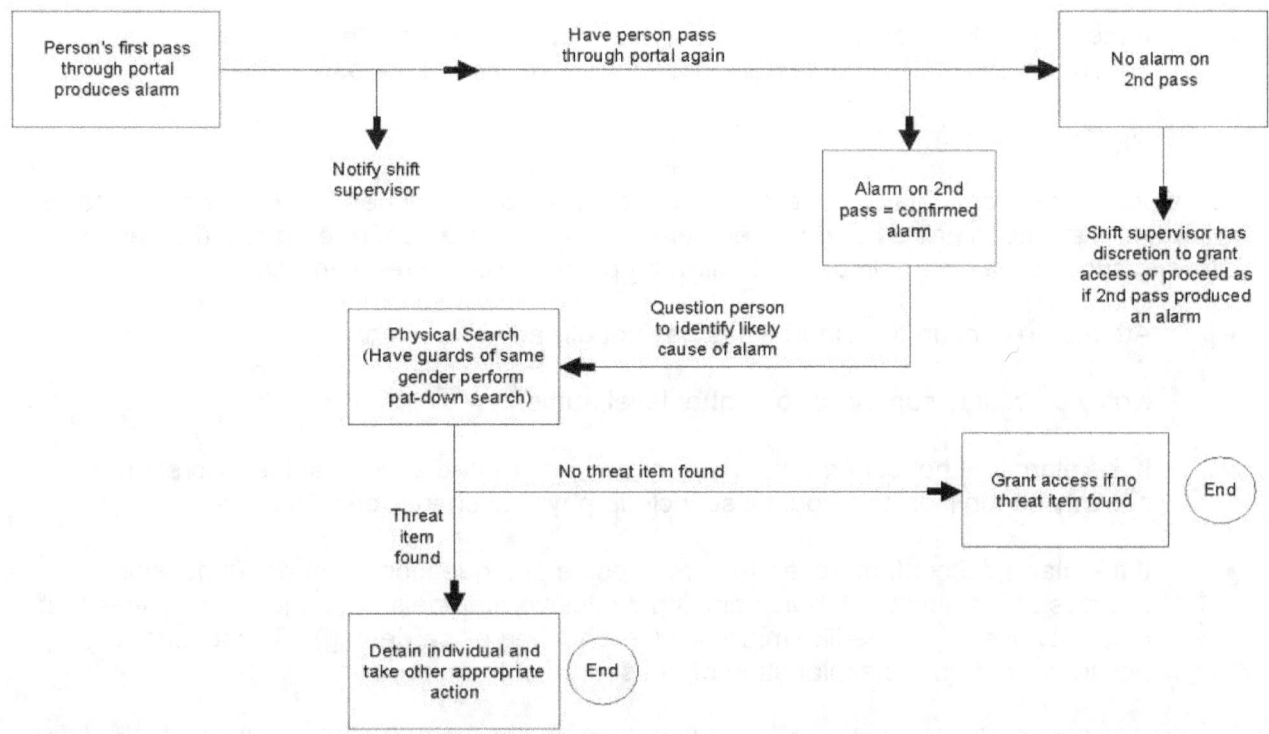

Figure 16: Sample decision tree for resolving an alarm from a personnel portal that detects trace explosive material

In all cases in which explosive trace detection alarms are recorded, the conditions must be documented. Recorded information should include the date and time of the alarm, the person involved, the type of explosive apparently detected, and the steps followed to resolve the alarm.

3.6.6.1 Alarm Resolution for Hand-Carried Objects

If a hand-carried object (such as a briefcase) produces an alarm, the following general procedures are recommended:

- Sample the object again to confirm the alarm after running at least two blank samples. Even if the alarm does not confirm, follow the steps below.

- Inform a security supervisor or higher level authority about the alarm.

- Perform a careful physical search of the hand-carried object being screened. Be sure to check all compartments where a piece of explosive material might be concealed. Asking the person carrying the object to open all compartments of the item is a best practice.

- If the alarm did not confirm and if the physical search is negative, the object may be taken through the checkpoint with the approval of the security supervisor.

- If the alarm did confirm, but the physical search of the object is still negative, the person carrying the item should undergo further screening (see alarm resolution for personnel).

- If the physical search identifies a suspected bomb or suspected explosive material, detain the person and contact security supervisor and LLEA as applicable.

3.6.6.2 Alarm Resolution for Personnel

If a person produces an alarm while passing through a portal or when having his or her hands swiped, or if a person has a hand-carried item that produces a confirmed alarm (i.e., alarms following two separate samplings), the following procedures are recommended:

- Attempt to confirm the alarm if this has not already been done.

- Notify a security supervisor or higher level authority.

- If the alarm did not confirm, the person may be granted access at the discretion of the security supervisor (this could also include physical search prior to entry).

- If the alarm did confirm, take the person aside and question them about possible sources of the alarm, such as handling explosive materials on the job or activities that might involve explosive-like materials (i.e., hunting or gardening). To the extent possible, confirm the explanation of the source of the alarm.

- Whether or not the likely source of the alarm can be identified, the individual should be subjected to a physical search to ensure the absence of explosives.

3.6.6.3 Alarm Resolution for Vehicles

If a vehicle being screened with trace explosive equipment produces an alarm, the following procedures are recommended:

- Attempt to confirm the alarm.

- Notify a security supervisor or higher level authority. This person may grant access if the alarm did not confirm. If the alarm did confirm, proceed as follows:

- Deny the vehicle access to the secure area and have the driver back out of the sally port or vehicle control point.

- Take the driver aside and question him or her about possible sources of the alarm. To the extent possible, confirm any explanation given of the source of the alarm.

- Collect and analyze more trace samples from other vehicle locations, including the passenger cabin and cargo hold.

- Clear the alarm by either conducting a very thorough physical search of the vehicle and its contents or by having a canine screen the vehicle, or both.

- Allow the vehicle access to the area only after the physical search verifies the absence of explosives and after the approval of the security supervisor has been obtained. Even then, having the vehicle escorted by security personnel is recommended.

3.6.7 Maintenance of Trace Detectors

The manufacturer's user manual should describe in detail the maintenance of commercial trace detectors, and maintenance activities and the frequency of performing them should be aligned with the manufacturer's recommendations. Like calibration training, maintenance training should be a part of the standard training that the manufacturer or vendor provides to the facility as part of the delivery and installation. Some maintenance operations can be performed routinely on site, but others may require a site visit by the manufacturer or vendor or returning the system to the manufacturer.

Some types of maintenance that may need to be performed on IMS-based systems include the following:

- Clean the instrument after it has become contaminated by a large detection from a sample that contained a large amount of trace explosive material. External contamination can often be removed by cleaning surfaces with wipes that have been wet with alcohol. Internal cleaning may require "baking out" the IMS instrument at a temperature well above the normal operating temperature for an extended period of time.

- Remove and clean the metal collection screens in the preconcentrators of personnel portals.

- Replace air purification cartridges (such as Dryerite cartridges).

- Maintain a stock of the clean sampling media (e.g., swabs) provided by the manufacturer.

- Replace the batteries or recharge the battery pack for certain hand-held instruments.

Since all maintenance is instrument dependent, this document cannot provide greater detail. When in doubt, consult the equipment manufacturer regarding maintenance procedures.

3.7 Explosives Detection Utilizing Canines

This section provides information on the testing of canines that are trained to detect explosive materials. As biological sensors, dogs represent a special case among the detection technologies discussed in the remainder of this section. Compared to manufactured trace explosive detection systems, dogs are more mobile and much better able to follow a scent rapidly to its source. Dogs excel at applications with a significant search component, including large vehicle and building searches. The main drawback of canines is their limited duty cycle. The length of time a dog can work without a break depends on the dog's training regimen; this time is typically limited to a few hours. A manufactured sensor, such as an IMS, can be operated on a 24/7 basis as long as someone can operate it. Dogs and manufactured trace sensors have several complementary capabilities and thus using both can be a best practice when not prohibited by cost. For example, a dog could be used to locate a suspected bomb in a building or vehicle, and the manufactured sensor could be used to determine what type of explosive is present.

The following discussion of canine testing and training is divided into the categories of initial training, certification training, and maintenance training, rather than the standard categories of calibration, operability testing, and performance testing used in describing testing of manmade equipment. Other sources sometimes use the term "performance testing" to describe some aspects of canine testing and training, but this document considers such testing to be part of either certification training or (more often) maintenance training.

To acquire a canine detection capability, facilities may develop the capability on site or rent the service from a canine service provider. Service rental is usually easier and more economical because the individual facility will not need to develop expertise in all areas of canine handling and detection. For example, the site can rely on the service provider for kenneling and healthcare expertise. Other Government or law enforcement agencies should be consulted to identify service providers whose dogs have performed acceptably in the past.

3.7.1 Test Objects for Canines

Key points include the following:

- Train dogs using bulk samples of the actual explosive materials the dogs are used to detect. The requirement to handle bulk explosive samples will place safety limits on where, when, and how the training can be performed. If a facility rents its canine detection capability, the service provider will probably already have training facilities set up to accommodate this training off site.

- The explosives RDX, PETN, TNT, dynamite, black powder, and double-base smokeless powder are considered mandatory substances for all explosive-detecting canines. A number of other explosives are listed as optional.

- The minimum weight of the explosive substance used for a training aid should be one quarter of a pound (113.4 grams). Larger masses should also be used in training, and the size used will depend on site-specific considerations both at the facility to be protected and (if different) the training site.

- As bulk explosives, training aids must be stored in accordance with all local, State, and Federal regulations.

- All training aids should be labeled. The information on the label should include (but is not necessarily limited to) the mass and type of explosive and other features that identify the training aid, such as a tracking code and emergency contact information.

- A material safety data sheet should be available for the explosive material used in a training aid. These documents can be obtained from a variety of sources, including manufacturers of explosives and at Web sites such as www.MSDSonline.com.

- To the extent possible, training aids should be stored in a manner that prevents cross-contamination of odors. All training aids should be stored in individual, airtight containers that are never opened within a storage bunker, and if possible, training aids containing volatile species (such as NG and EGDN) should be stored in separate bunkers from other training aids. Frequent replacement of training aids (at least once every 12 months) is another means of limiting the possible effects of cross-contamination. A training aid should be replaced sooner if it is known to be contaminated.

3.7.2 Initial Training of Canines

Initial training of a canine is designed to ready the dog for certification training. Rather than developing an in-house training program, it is probably most cost effective for many sites to contract with a service provider that has already trained and certified its dogs. The provider should possess a record of past performance through canine detection services to other Government entities.

Initial training typically includes varying quantities of the different explosives, with masses typically varying by an order of magnitude or more. The training should represent all conditions that may be encountered in certification testing. The canine should be exposed to training aids placed at different heights and depths, as defined in the certification training procedures. All training should include a variety of blank searches in which no test object is present to demonstrate that the dog responds only to legitimate threat materials. The length of time required for initial training varies for each dog.

Initial training has two possible outcomes:

- The dog progresses so it can pass rigorous certification training and proceed to deployment.

- The dog is eventually deemed not certifiable because of substandard performance and is never deployed.

3.7.3 Certification Training of Canines

Certification training involves rigorous evaluation of a dog's ability to detect explosives and represents a critical decision point with regard to the dog's deployment. If a dog passes certification training it can be deployed, always with the requirement that additional maintenance training will be performed on a regular basis. If a dog cannot pass certification training, it cannot be deployed. Note that the canine and its dedicated handler will be certified as a team. If a facility rents a canine capability from a vendor, the canine/handler team obtained may already have undergone documented certification training and the facility should ask to receive and review this documentation. However, consultation should begin well in advance of deployment, since a dog's certification training may need to be changed or supplemented to accommodate site-specific operational conditions.

3.7.4 Maintenance Training of Canines

Maintenance training is intended to sustain the performance of a canine/handler team at a high level over the dog's working lifetime after certification training has been completed. The maintenance training should include scenarios representing all of the dog's working conditions. It should also include vital aspects of canine/handler performance, such as leash control, canine obedience, and search patterns. All maintenance training must be carefully documented, whether it is performed on or off the site where the dog is deployed. When contracting with a service provider, the site should have full access to all training and certification records. A canine/handler team is typically required to spend an average of 4 hours per week training to maintain detection capabilities.

If certification training was performed at a location other than the facility at which the canine will be deployed, it is strongly recommended that some follow-on training at that facility be performed before deployment of the canine/handler team at the facility. Such training needs to be site specific because of site environmental factors and the limitations that may exist in handling explosive materials at the site. To the extent possible, site training should include tests similar to at least some tests carried out in certification training, but in an environment that is as close as possible to the actual working environment of the canine at that site. This initial onsite training is sometimes referred to as performance testing, and proper documentation of the results is extremely important.

In addition to weekly training by the canine/handler team alone, supervised training should be conducted periodically by a qualified trainer other than the handler to identify and correct deficiencies and perform proficiency assessments.

Maintenance training should also include regular blind challenge tests where, unknown to the handler, a test object that the dog trains on has been concealed inside a vehicle or other item being screened. Regular success in these blind challenges will maintain a high level of confidence that the dog is performing as desired.

3.7.5 False/Nuisance Alarms and Alarm Resolution for Canines

3.7.5.1 False Alarms

As noted previously, the dog's alarm rate in a realistic working environment with no threat article present should not exceed 10 percent. In most cases, the false alarm rate should be well below 10 percent for a well-trained dog, which is an operational necessity given the time and effort needed to clear false alarms.

3.7.5.2 Alarm Resolution for Canines

When a dog indicates the presence of a threat material, great care must be taken in resolving the alarm. Since different facilities have different resources available for alarm resolution, alarm resolution can be site specific, depending on these resources and on the circumstances of the alarm; however, alarm resolution should always follow procedures that have been thought through and documented in advance. An example of an appropriate set of alarm resolution procedures is provided below.

The object that appears to be producing the alarm should not be moved or even touched except by qualified bomb squad personnel and even then only after all other possible analysis of the item has been performed. The following procedures are recommended:

- Move personnel away from the possible threat object and have one or more security officers enforce a standoff distance.

- Notify a security supervisor or higher level authority of the alarm.

- Notify LLEA as appropriate and await their arrival.

- If a trace detection system is available, attempt to collect a trace sample without touching the item and analyze this to see if any explosive material is detected and what type. Noncontact trace sampling can involve collecting an air sample by vacuuming near the object or collecting a swipe sample from the ground, the floor of the vehicle cargo hold, or whatever other surface the object is resting on and immediately adjacent to the object.

- When LLEA bomb squad personnel arrive, turn the situation over to them for final resolution and disposal of the object as required.

This rather time-consuming process indicates just how costly false canine alarms can be. However, experience shows that the rate of false alarms is extremely low for well-trained canines.

4. SCHEMATIC DRAWINGS OF ACCESS CONTROL PORTALS

This section contains schematic drawings of the following:

- a personnel access control portal for a protected area

- a vehicle access control portal for a protected area

These drawings represent a means of setting up the access control portals and associated screening technologies for effective access control. Other checkpoint layouts may also be appropriate as long as all required screening is performed.

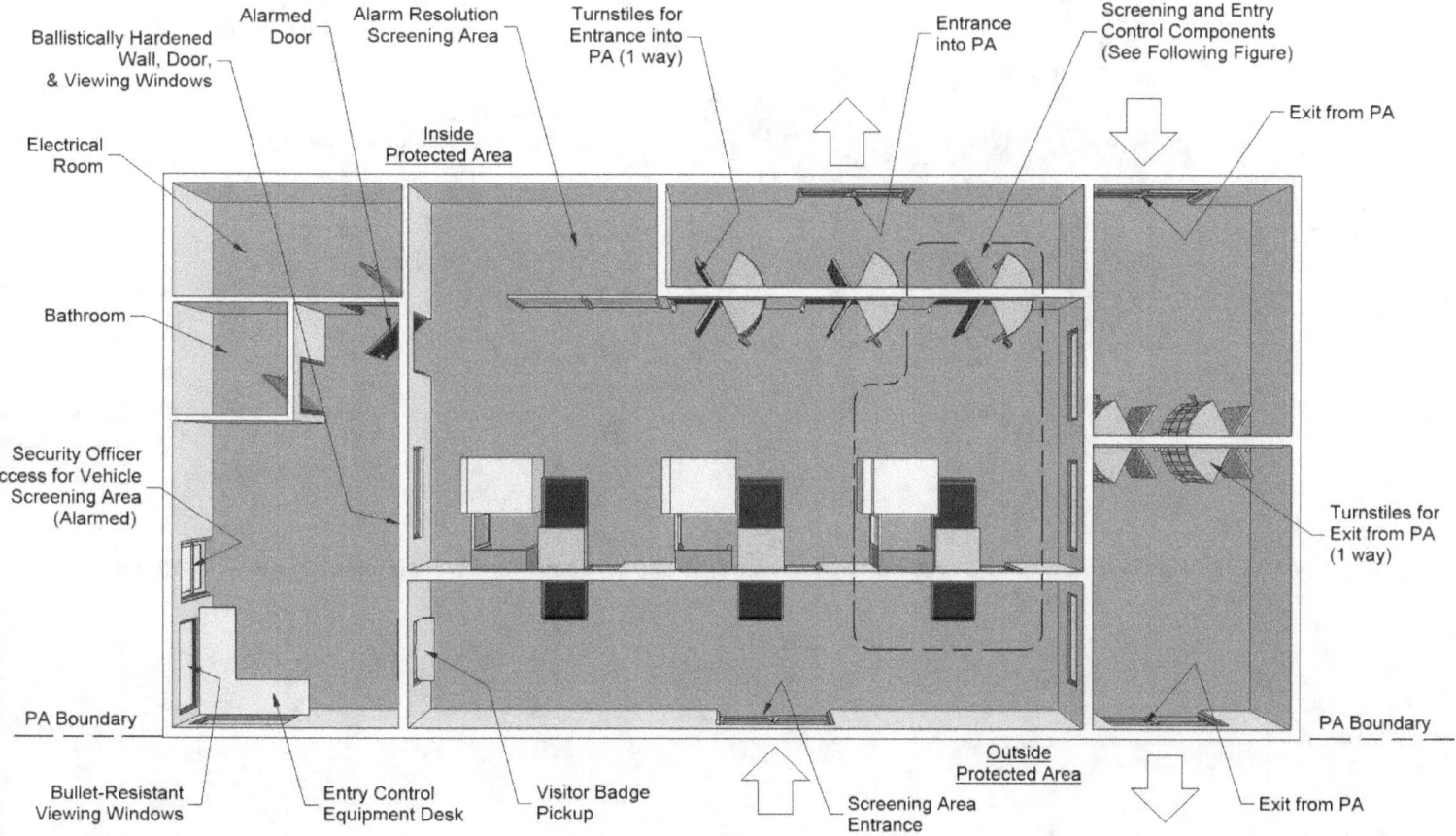

Figure 17: Personnel access control portal for a protected area

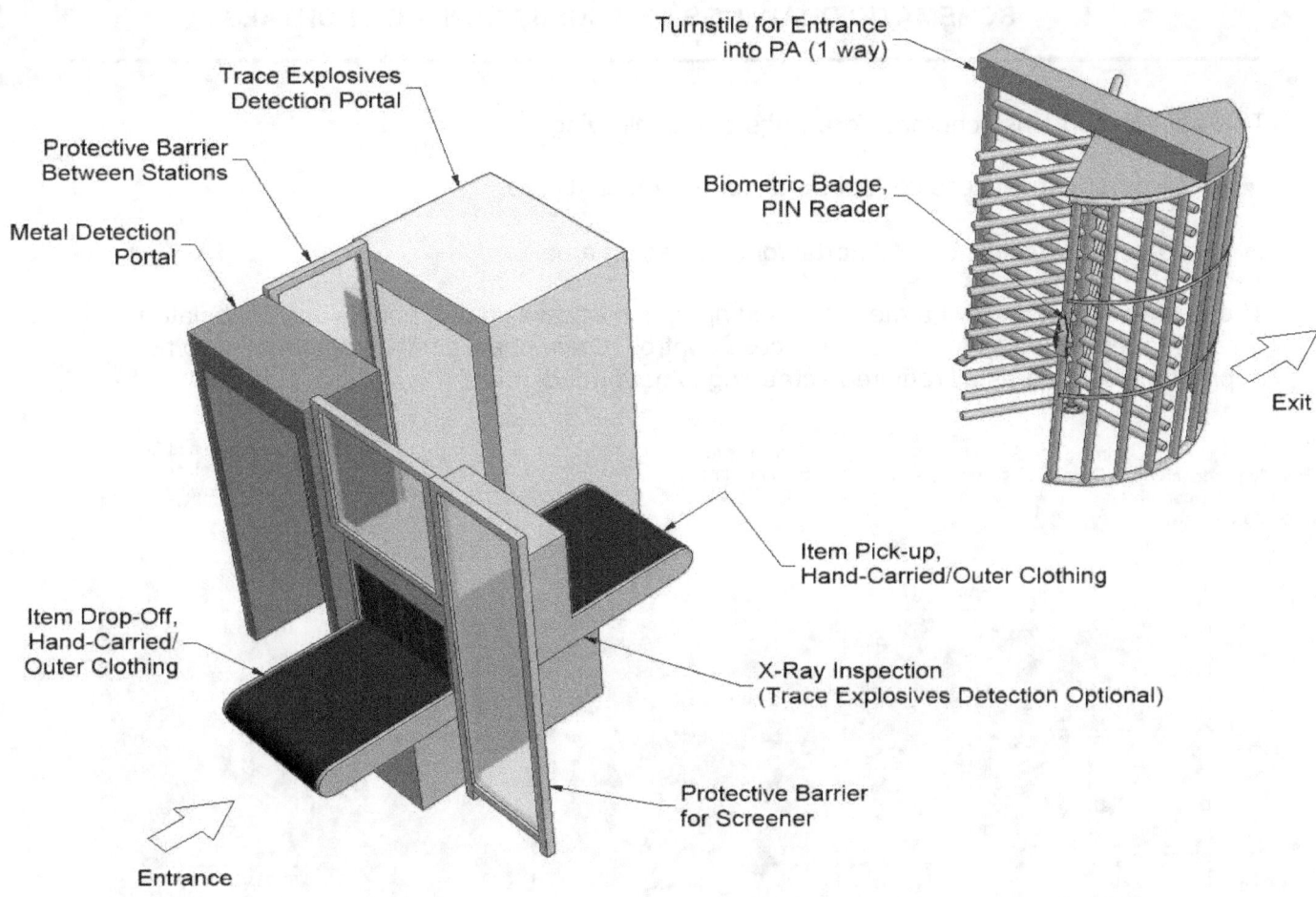

Trace Explosives
Detection Portal

Protective Barrier
Between Stations

Metal Detection
Portal

Turnstile for Entrance
into PA (1 way)

Biometric Badge,
PIN Reader

Exit

Item Drop-Off,
Hand-Carried/
Outer Clothing

Entrance

Item Pick-up,
Hand-Carried/Outer Clothing

X-Ray Inspection
(Trace Explosives Detection Optional)

Protective Barrier
for Screener

Figure 18: Detailed view of one of the three search trains shown above for screening personnel entry to the protected area

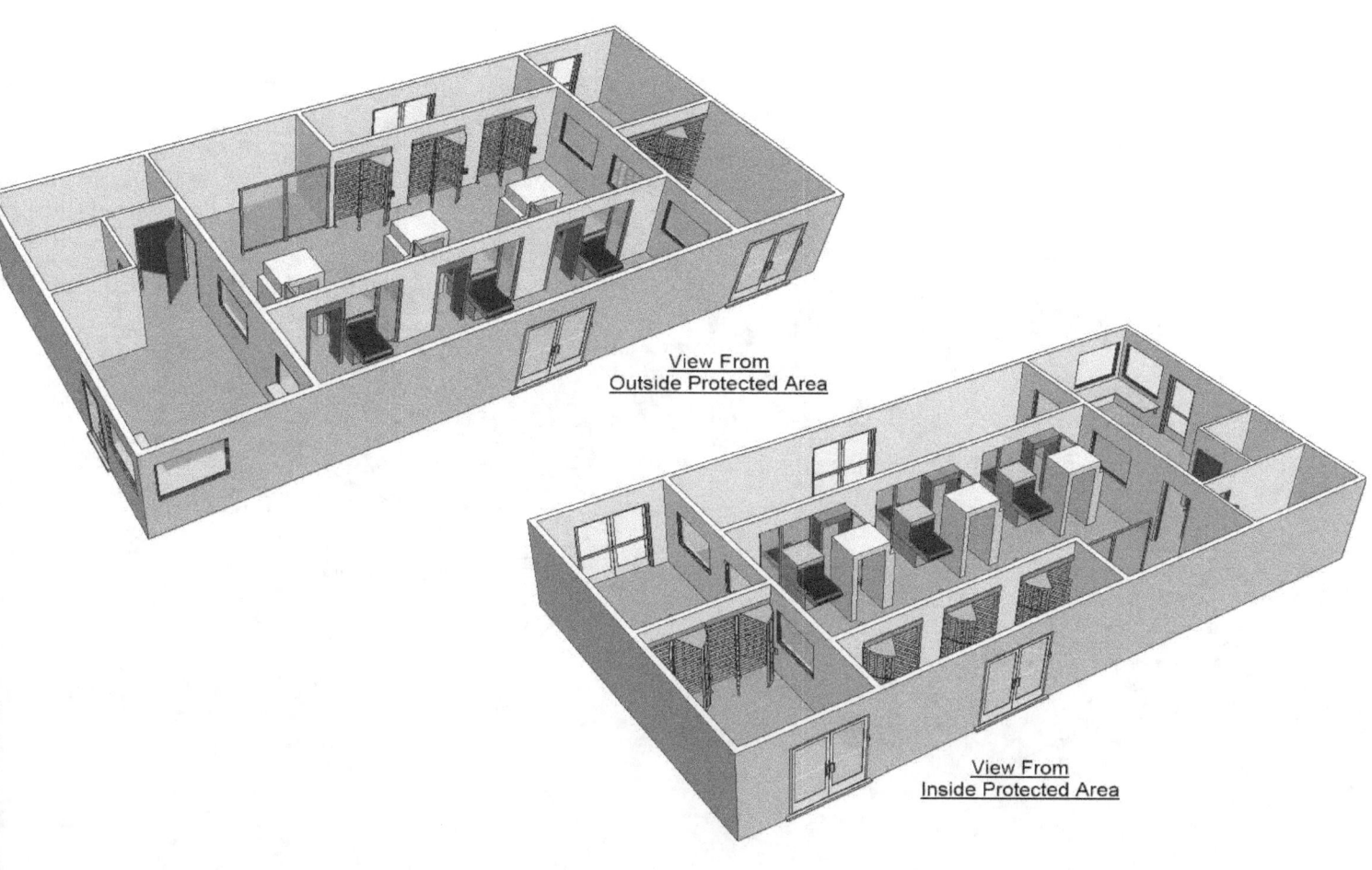

View From
Outside Protected Area

View From
Inside Protected Area

Figure 19: The protected area personnel access control portal viewed from outside and inside the protected area

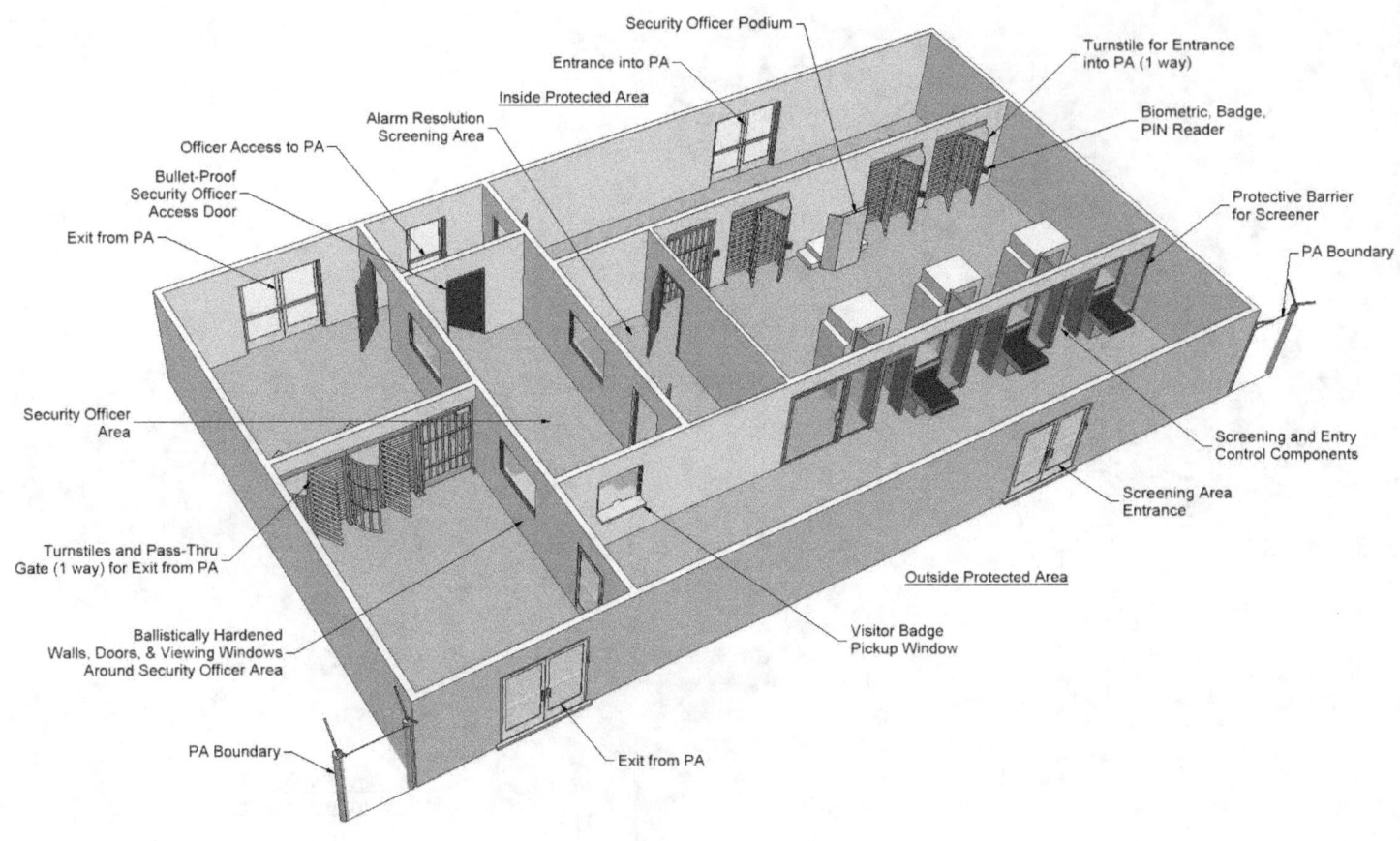

Figure 20: Another acceptable layout for a protected area personnel access control portal

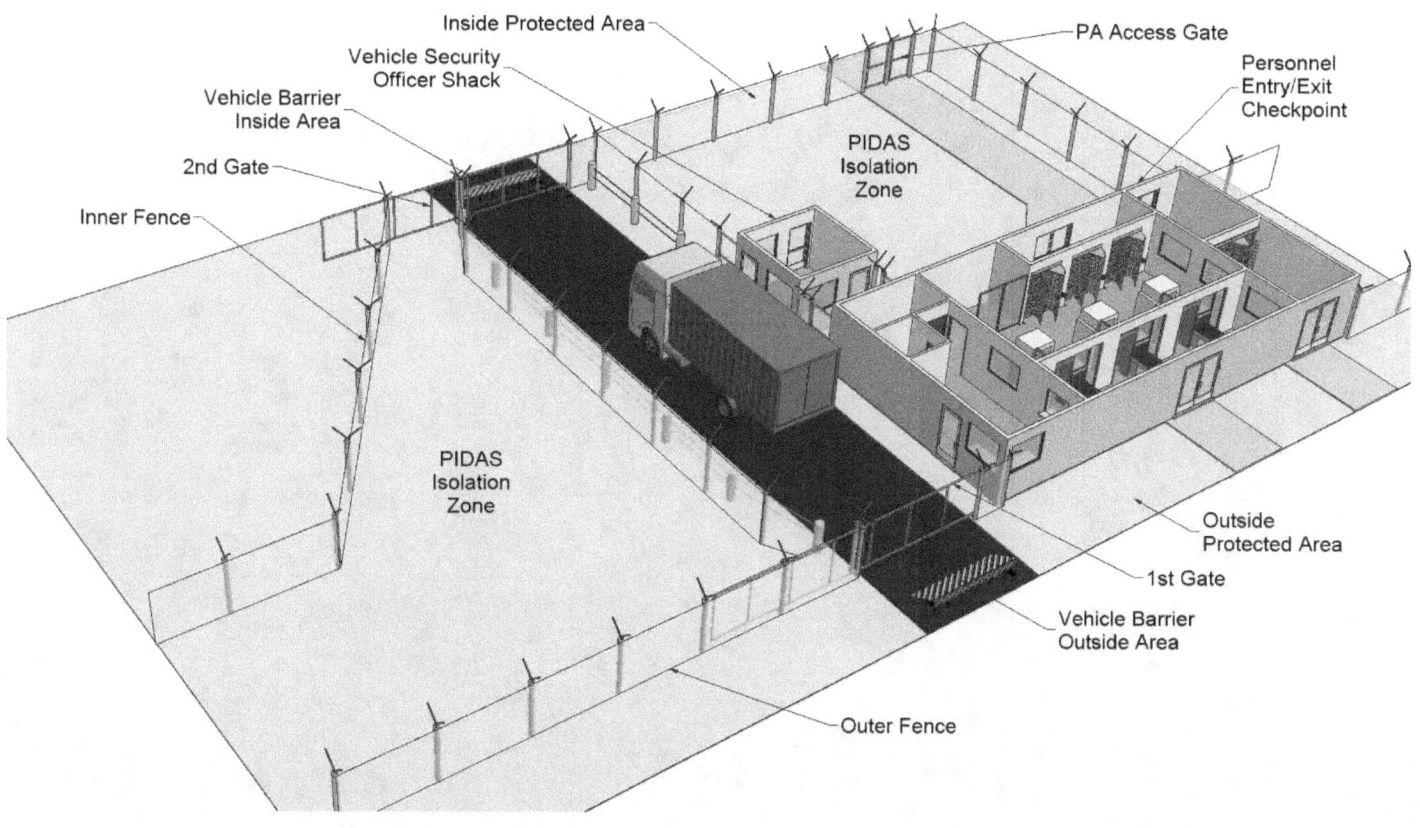

Figure 21: Vehicle access control portal (sally port) for a protected area. The sally port is shown adjacent to the personnel access control portal that was illustrated in the preceding figures.

NRC FORM 335
(12-2010)
NRCMD 3.7

U.S. NUCLEAR REGULATORY COMMISSION

BIBLIOGRAPHIC DATA SHEET

(See instructions on the reverse)

1. REPORT NUMBER
(Assigned by NRC, Add Vol., Supp., Rev., and Addendum Numbers, if any.)

NUREG-1964

2. TITLE AND SUBTITLE

Access Control Systems

3. DATE REPORT PUBLISHED

MONTH	YEAR
April	2011

4. FIN OR GRANT NUMBER

5. AUTHOR(S)

Division of Security Policy, Office of Nuclear Security and Incident Response

6. TYPE OF REPORT

7. PERIOD COVERED *(Inclusive Dates)*

8. PERFORMING ORGANIZATION - NAME AND ADDRESS *(If NRC, provide Division, Office or Region, U.S. Nuclear Regulatory Commission, and mailing address; if contractor, provide name and mailing address.)*

Division of Security Policy
Office of Nuclear Security and Incident Response
U.S. Nuclear Regulatory Commission
Washington, DC 20555-0001

9. SPONSORING ORGANIZATION - NAME AND ADDRESS *(If NRC, type "Same as above"; if contractor, provide NRC Division, Office or Region, U.S. Nuclear Regulatory Commission, and mailing address.)*

Same as above

10. SUPPLEMENTARY NOTES

11. ABSTRACT *(200 words or less)*

This report provides technical details applicable to access control methods and technologies used to protect facilities licensed by the U.S. Nuclear Regulatory Commission. It contains information on the application, use, function, installation, maintenance, and testing parameters for access control and search equipment and the implementation of protective measures that support access control. This information is intended to assist licensees in designing, installing, employing, and maintaining access control systems at their facilities.

12. KEY WORDS/DESCRIPTORS *(List words or phrases that will assist researchers in locating the report.)*

Access Control Systems

13. AVAILABILITY STATEMENT

unlimited

14. SECURITY CLASSIFICATION

(This Page)

unclassified

(This Report)

unclassified

15. NUMBER OF PAGES

16. PRICE

NRC FORM 335 (12-2010)

Printed
on recycled
paper

Federal Recycling Program

NUREG-1964

Access Control Systems

April 2011